Understanding WAP

Wireless Applications, Devices, and Services

For a listing of recent titles in the *Artech House
Mobile Communications Library*, turn to the back of this book.

Understanding WAP

Wireless Applications, Devices, and Services

Marcel van der Heijden
Marcus Taylor
Editors

Artech House
Boston • London
www.artechhouse.com

Library of Congress Cataloging-in-Publication Data
Understanding WAP: wireless applications, devices, and services / Marcel van der Heijden, Marcus Taylor, editors.
    p.   cm. — (Artech House mobile communications library)
    Includes bibliographical references and index.
    ISBN 1-58053-093-1 (alk. paper)
    1. Wireless communication systems.   I. Heijden, Marcel van der.
    II. Taylor, Marcus.   III. Artech House telecommunications library.
TK5103.2.U53   2000
621.3845—dc21   00-030611
                                         CIP

British Library Cataloguing in Publication Data
Understanding WAP: wireless applications, devices, and services. — (Artech House mobile communications library)
    1. Wireless communication systems   2. Computer network protocols
    I. Heijden, Marcel van der   II. Taylor, Marcus
    621.3'845
    ISBN   1-58053-093-1

Cover and text design by Darrell Judd

International Standard Book Number: 1-58053-093-1
Library of Congress Catalog Card Number: 00-030611

10 9 8 7 6 5 4 3

Contents

Preface

Over the last five years or so a revolution has taken place, and continues to do so, that is changing the way we communicate and interact with each other and our environment. The advent of the Internet and the ubiquitous presence of mobile personal communications systems are the most prominent examples of this. The Internet has made it possible to exchange information instantly on a worldwide scale and has released unprecedented amounts of information into the public domain. Mobile communications allowed personal communications on an anytime, anywhere basis.

While mobile access is possible using standard technology, more is needed to make mobile access truly seamless. The creation of a world standard for mobile Internet access and information access would ensure a rapid growth and a basis for all mobile communications users to interact. The wireless application protocol (WAP) is an attempt to do just that, and judging by the industry support at the time of this writing, a successful one.

The impact of WAP on all parties in the value chain of services delivery and information exchange is hard to underestimate. WAP will be an additional channel for some to offer their services, a new enabler for others who can now offer services that could not be offered before. It will be a treat for some and an opportunity to others, depending on the approach taken. Most of all, it will release an enormous potential of services and applications for the end users. In many ways WAP will probably change the world as the Internet and mobile communications have done before.

Understanding WAP: Wireless Applications, Devices, and Services is a comprehensive description of many aspects of WAP. It will explain some technicalities of WAP and the system elements involved for a full understanding of the potential of WAP, though it is not specifically geared towards engineers only. It will also describe the impact of WAP on wireless operators and service providers. This book is aimed at engineers and technical managers as an introduction to WAP and at business managers as a way to gain an understanding of the technical background of WAP.

Finally, the editors wish to express their gratitude to all who have contributed to the creation of this book and who have made working on this book an interesting experience.

Marcel van der Heijden and Marcus Taylor
Editors

CHAPTER

1

Contents

Introducing the Wireless Application Protocol

Per Ocklind

1.1 Introduction

Today, the wireless network is mainly used for voice communication, where voice mail might be the most popular value-added service, if any. However, a new buzzword is increasingly being mentioned in the marketplace: WAP, or wireless application protocol.

WAP is a completely new concept. It provides data-oriented (nonvoice) services to the mass market and is capable of being beneficial, anywhere and anytime, to far more end users than the personal computer. WAP is a global standard that is independent of the underlying bearer. With WAP, a new dimension will be added to the use of mobile phones, through the introduction of new data-oriented mass-market services.

1

1.2 How it all started

In mid-1997 Ericsson, Motorola, Nokia, and Phone.com (previously Unwired Planet) came to an agreement to mutually define a new protocol for mobile devices. The objective was to offer new wireless datacommunication services to end users, both in the form of telecommunication-related and Internet-oriented applications.

1.2.1 Why was this done together?

Over the years, several protocols have been defined by various players in the market, for various types of applications. In the beginning Unwired Planet had its handheld device markup language (HDML), a protocol for Internet access to be used over cellular digital packet data (CDPD) networks. At the CeBIT exhibition in 1997, Nokia launched a protocol called tagged text markup language (TTML), a protocol with a similar focus as the HDML protocol, but designed to be used in the GSM world. Ericsson, in turn, was in the process of launching a protocol mainly focusing on telecommunications-related and messaging applications (e.g., call handling and call control) to be used inside the GSM networks called intelligent terminal transfer protocol (ITTP).

These three protocols represented only a fraction of the different protocols defined by different organizations and available in the marketplace. This fragmentation prevented the market for wireless applications from taking off. In order to clean up this fragmentation, forces were joined in defining a common platform, a common protocol, embracing Internet access, messaging (see Chapter 9) and telecommunications related applications, and everything in between (see Figure 1.1).

It was realized that if this common platform was not created, and proprietary projects and solutions continued to be developed, the consequence would be that the market would take up an unresponsive attitude to these projects; it would remain in an idle mode.

The first joint meeting took place in Seattle in June 1997 and was followed by quite an extensive meeting schedule. The intention from the start was to broaden the group of companies working with WAP. However, as all parties involved felt that it was very important to get a first draft specification out as soon as possible, it was stated that the doors for new members would be opened only when the first draft was released and available.

Why a new standard?

The solution:

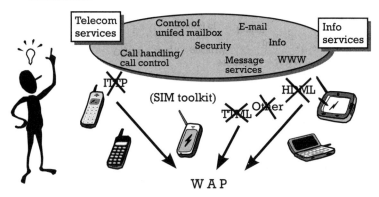

Figure 1.1 Schematic overview of how the different services and enabling technologies are related and replaced by WAP.

1.2.2 A worldwide solution

Since one of the main goals was to make this protocol global and generic, we stressed that it should be independent of the underlying bearer. That is, it should operate on any cellular bearer (e.g., GSM, D-AMPS, CDMA, and PDC) and even on nonvoice systems, such as Mobitex and paging systems.

1.2.3 The WAP Forum Ltd.

In December 1997, the four parties formed a company, WAP Forum Ltd., to control and manage the continued development of WAP. The first draft of the WAP 1.0 specifications was released in April 1998, and consequently the door for new members to enter the WAP Forum was opened. In June 1998, one year after the first joint meeting, the WAP Forum had its first meeting with the new member constellation. The number of members at this time was approximately 40.

1.2.4 The first commercial version of WAP

In May 1999, 11 months after the release of the WAP 1.0 specifications (and after a large number of working meetings), the first commercial version of WAP (WAP 1.1) was finalized. The number of members has increased steadily, and the present figure (May 2000) is now more than 200 members. By now, many implementations of WAP components are being introduced and an enormous diversity of WAP services have been rolled out at an impressive pace.

1.3 A brief description of what it is all about

The following can be seen as an executive summary of WAP in some detail from a technical standpoint. These components will be described in more detail in separate chapters of this book.

1.3.1 Optimized for wireless communications

In addition to providing end users with new services, WAP has been designed for the economical use of the resources available in the telecommunications network. Over-the-air interface communication is binary coded to use the bearer services as efficiently as possible. Message headers and parts of messages in plaintext format that are frequently sent are represented as bytes. The original content is then restored in the receiver (the WAP browser).

In the wireless world, the handset also sets some limitations such as display size, number of keys, CPU capacity, etc. This means that the applications will differ from the ones we are using for normal Internet surfing.

1.3.2 Deck of cards

The handling of services is based on a deck of cards metaphor. A deck is sent from the network to a user's terminal when he or she enters a command to invoke a service. The user can then navigate through the complete deck to make a choice. If the desired choice is not included in the deck, another deck can be requested by command. When the user has made his or her choice and entered the relevant command, the requested action is performed or information is retrieved. Depending on the capacity of the mobile phone, the decks and cards can be cached in the WAP phone (terminal) for future use.

1.3.3 The WAP stack

One of the objectives of specifying the WAP was to make the mobile phone a first-class citizen of the Internet. Therefore, it was only natural that an Internet-oriented approach be adopted. As Figure 1.2 shows, the WAP stack is similar to the layers used in the Internet.

The following entities are defined in WAP.

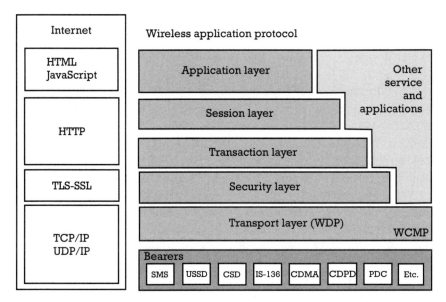

Figure 1.2   Schematic representation of the WAP stack of protocols (right) and entities and the Internet protocol stack (left). The two protocol stacks are similar in many ways.

1.3.3.1   Microbrowser

The microbrowser can be compared to a standard Internet browser, for example, Netscape Navigator or Microsoft Internet Explorer. The applications we are accessing must be written in the new markup language defined within WAP, which is called wireless markup language (WML). WML, however, is structured rather similarly to HTML (WML follows XML as opposed to HTML) and can easily be converted from it. The conversion (or filtering) of existing HTML content to WML is not covered within the WAP specification. The same applies to the actual implementation of a browser. The WAP specification describes the minimal functionality that they should contain.

1.3.3.2   WMLScript

The wireless markup language script (WMLScript) is a scripting language very similar to JavaScript/ECMAScript. It provides a means of reducing airtime by enhancing the capability of the handset; that is, it enables the handset to process more information locally before sending it to the server.

Chapter 2 describes in more detail the wireless application environment (WAE), of which WML and WMLScript are parts, from a developer's perspective.

1.3.3.3   Wireless telephony application and the WTA interface

The telephony part of WAP is called the wireless telephony application (WTA). A WTA application is using the wireless telephony application interface (WTAI) in order to create call-control and call-handling applications, for example, the definition of call chains and various options when a call is received, integrated within the visual WAP user environment. The WTAI ensures that the WTA user agent can interact with mobile network functions (e.g., setting up calls) and device-specific features (e.g., phonebook manipulation). For further information, refer to Chapter 4, where WTA and WTAI are described in more detail.

1.3.3.4   Content formats

The content formats that are supported in WAP include (among others) business cards (vCard) [1], calendar events (vCalendar) [2], and so on. Thus, existing standards and technologies are leveraged and can be easily migrated into WAP applications and services.

1.3.3.5   A layered telecommunication stack

Includes transport, security (see Chapter 7), and session layers.

1.3.3.6   The WAP gateway

In order to reach the Internet world, the WAP-enabled phones must travel via a WAP gateway (G/W). The WAP G/W acts as an intermediary, connecting the mobile network and the Internet by translating the hypertext transfer protocol (HTTP) to the wireless session protocol (WSP). This is often referred to as protocol translation.

For general WAP services, the operator offering generic access to WAP applications on the Internet will typically host the WAP gateway. However, some companies or organizations might want to let their employees access intranet information (i.e., information inside the firewall). In this case, the company will have a WAP G/W of its own.

Figure 1.3 gives a simplified picture of the different WAP entities within the wireless network.

WAP gateways or proxies may also perform a conversion of the content that is being requested by the WAP client. A WAP browser may, for instance, request an HTML page, which is then converted into WML by

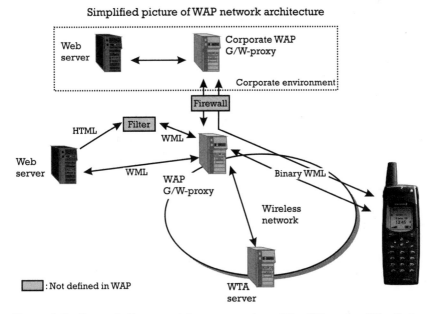

Figure 1.3 Network diagram giving an overview of the different entities that may exist in a wireless network that supports WAP.

the WAP gateway. Similarly, a plaintext WML file that is requested may be encoded into the binary format before it is sent to the WAP client. Chapter 5 describes in more detail WAP gateways and how they can be employed.

1.4 Applications using WAP

With WAP a user with a mobile phone supporting WAP will have access to information that is available on the Internet in WML and WMLScript format as well as possible telephony applications provided by the operator. WAP encompasses almost the entire wireless community, making it ubiquitously available to mobile communications users. This suggests that (besides the physical characteristics of the WAP-enabled device) only the imagination sets the limit to what is possible with WAP. Listed here are a few examples of possible applications.

1.4.1 Information retrieval on the Internet

WAP can be employed to reach information on the Internet. However, the WAP browser cannot be used exactly in the same way as an ordinary "surfing tool," since the mobile phone sets some limits on input and output capability, memory size, and so forth. As the user experience will be of paramount importance for a broad take-up of WAP services, Chapter 3 discusses WAP applications from a usability and user interface design perspective.

1.4.2   The serviceperson application

With a WAP-enabled mobile phone, servicepeople on duty can access their company inventory to check whether or not a spare part is available and directly inform customers about the situation. Of course, they can use the same application to order spare parts, and will immediately receive a confirmed delivery date. When a job is done, the service team members can sign off and report the current status of the site.

1.4.3   Notification applications

By means of agents residing in servers, users can be notified of e-mail and voice mail messages that have been sent to them. They can interactively request that more information be sent to their phones, or order a printout on a fax machine of their choice.

Users interested in buying or selling shares can define a buying or selling profile that shows, for example, what stocks they are interested in, and at what quote they might be willing to buy or sell. When a specific quotation has passed a defined trigger value, the agent notifies them and asks if they want to make a transaction. Such notifications can be "pushed" to the mobile WAP user using the push facilities provided by the WAP specifications. Chapter 6 describes the push architecture and implementation issues in considerable detail.

1.4.4   Mobile electronic commerce

Users can have access to payment services for bank transactions, ticket offices and wagering systems, and electronic auctions. The role of WAP in mobile financial services is discussed extensively in Chapter 10.

1.4.5   Telephony applications

A user can have access to services that handle call setup, in combination with other services provided by a wireless operator. A typical example involves a menu (see Figure 1.4), defined by the user, which is displayed for each incoming call. This menu allows the user to decide whether to answer or reject the call, or to forward it to another extension or to a voice mail service. The WTA part of the WAP specifications is extensively described in Chapter 4.

1.5 Benefits and market opportunities

WAP will add quite a few benefits and opportunities for all players on the wireless market. Listed here are a number of these opportunities.

1.5.1 Operator benefits

For the network operators, WAP means that they get a tool by which they can offer a new category of services to end users. They can quickly create unique and innovative services and provide access to third-party services that are available on the Internet. Operators can improve their customer services and help desks with a reduced cost by providing access to information residing on their networks. With the introduction of WAP, they will also remotely be able to tailor the menus and the interface of customers' telephones to further differentiate their services. In many respects, WAP will be a tool for operators to improve their services and thus reduce churn. Chapter 8 describes in detail the issues that arise around the deployment of WAP by operators.

Figure 1.4 Typical example of a text-based WAP-enabled phone where WAP is used to create an interface to the phone's functionality.

1.5.2   Content provider benefits

WAP allows the content provider to aggregate, create, and broker content that can be accessed from practically anywhere using the WAP distribution channel and interface. This content, which is available on a server, can be accessed from any wireless device almost anywhere on the globe. The current trend of managing content in XML [3] format, with the aim of effectively separating the content from the media format used for distribution, also allows for the distribution of this content through a number of different channels such as WAP, HTML, text-to-speech, interactive television, etc.

1.5.3   End user benefits

More and more people are using their PCs to retrieve information from global sources. Many users have found increased possibilities of executing services on their fixed phones around-the-clock using an interactive voice-response system. Now, even greater possibilities are facilitated through services based on WAP technology (see Figure 1.5). To make it practical to use a phone for accessing services, the user must be able to personalize the services, for example, from a Web page.

Thanks to WAP, end users can access the following services, and many more, by entering the relevant choice via their terminal:

- Banking;
- Stock exchange quotations;
- All types of news, such as sports, financial, etc.;
- Weather forecasts;
- Corporate applications such as inventory balance, enterprise resource planning applications, etc.;
- Tele-services;
- Games, gambling, and other entertainment applications;
- Geographic information.

1.6   Challenges to the network operator

In the world of new services and opportunities, network operators are likely to meet new challenges. They run high-quality networks with good coverage as a basis for their business, but they must also decide on the

Figure 1.5 General overview of the types of possible services based on WAP technology.

next step in developing their business: Should they remain pure network operators only, providing access to a multitude of new services, or should they take an active part in providing the services themselves? With WAP, operators can choose to open up access to all sites on the Internet, by allowing all universal resource locators (URLs) to be accessed by their subscribers, or they can choose to exercise full control, solely allowing access to those sites that are included in a service package.

1.6.1 Creating a service package

Given the similarity of the WAP programming model to that used for Internet applications development, it is relatively easy to develop WAP applications. Thus, wireless network operators create large parts of their WAP service portfolios themselves and bundle them in an attractive fashion, with third-party applications, for instance. It is doubtful whether wireless operators will actually choose to do so, however, since a wide variety of companies already offers sets of WAP services with attractive content. The other extreme is to outsource all services to a partner that handles the services, including subscriptions.

To keep control and yet allow access to content that is already available on the Internet, operators will negotiate directly with content providers, or with content brokers that provide an assortment of services. In addition, the operators will have to introduce a service-management system that allows end users to select the services they want.

1.7   What is next?

1.7.1 Interoperability

It is, of course, essential that all the WAP products produced by various suppliers can communicate with each other. In order to establish the interoperability between WAP devices and servers from multiple suppliers, the WAP Forum has created specifications for conformance requirements. A neutral company outside of the WAP Forum handles the day-to-day management of the certification process. Apart from this, the WAP Forum is running pair-wise interoperability tests between products from different companies.

1.7.2 Products, of course!

Most (if not all) of the big players in the telecom and datacommunications arena have joined the WAP Forum. It is quite clear that to date, WAP is supported by the large majority of the wireless community who are committed to making WAP a success. However, as always, compelling products and applications are required in order to get the ball rolling. The fact that many companies outside of the WAP Forum are also in the process of developing products and applications supporting WAP further strengthens WAP as a future profitable solution.

In 1999, several companies launched WAP products for WAP-enabled handsets (e.g., WAP browser software) and servers (e.g., WAP gateways) as well as WAP applications (e.g., WAP mobile banking applications). At present it is clear that the market started to move at the end of 1999, and we have seen a full takeoff in 2000.

1.7.3 Liaisons with other organizations

The market is constantly changing. This change will set new requirements on both the WAP specifications and their evolution. The WAP Forum needs to have an ear to the ground in order to prepare for these requirements and keep abreast of changes in the developer and standardization community. For this reason it is essential that the WAP

Forum strengthen relations with organizations with which it already has an established liaison (e.g., W3C, ETSI, IETF, TIA, ECMA, etc.) as well as continue to meet with other relevant organizations.

1.8 Conclusions

WAP enables network operators and content providers to reach a mass market. Although WAP-enabled phones are just in the process of being launched, there are today approximately 200 million GSM subscribers using wireless devices. Looking less than five years ahead, forecasts by major handset manufacturers predict this figure will probably rise to 700 or 800 million, and we have good reason to believe that the majority of handsets in use will support WAP. Without a doubt, WAP will add a new dimension to the use of mobile phones.

References

[1]   vCard—The Electronic Business Card, Version 2.1, The Internet Mail Consortium (IMC), Sept. 18, 1996, http://www.imc.org/pdi/vcard-21.doc.

[2]   vCalendar—The Electronic Calendaring and Scheduling Format, Version 1.0, The Internet Mail Consortium (IMC), Sept. 18, 1996, http://www.imc.org/pdi/vcal-10.doc.

[3]   St. Laurent, S., and E. Cerami, *Building XML Applications*, New York: McGraw-Hill, 1999.

CHAPTER

2

Contents

The Wireless Application Environment for Creating WAP Services and Applications

Marcel van der Heijden and Martin Frost

2.1 Introduction

In recent years, we have witnessed the incredible growth of the Internet: By now, almost all sources of information and many applications and services are available through this medium. Moreover, a whole new industry has arisen that offers networked applications and services that differ radically from traditional business models.

While the growth of the Internet has been phenomenal, with e-mail often mentioned as one of the "killer applications," there is one communications channel that has seen growth even more staggering. That channel currently has more than twice as many users as the Internet. We are talking about wireless personal communications.

15

While the Internet offers a relatively rich user experience and provides an enormous amount of information, its use is still largely limited to the computer at home or in the office. Wireless communication allows people to communicate regardless of where they may be, but unfortunately suffers some drawbacks when it comes to accessing information or delivering data services, an important one being the typically rather limited user interface capabilities of the wireless devices and low data rates of the wireless channel.

Bringing together the two technologies while at the same time maintaining the strong points of each can create a new revolution all over again. Handset manufacturers, network equipment providers, wireless operators (see Chapter 8), value-added service providers, content providers, e-commerce operators and banks (see Chapter 10), newspapers, the transportation industry, corporate IT departments, and many, many more understand that the combination of the Internet and the wireless world will have far-reaching implications for them. The result of this synergy is WAP.

WAP encompasses the specifications of a whole range of protocols and systems (see [1] and Chapter 1 for an overview). The part of WAP that developers will use to develop their wireless applications and services is the wireless application environment (WAE, see [2]). The WAE consists of three main parts:

> The wireless markup language (WML);

> The wireless markup language script (WMLScript);

> The wireless telephony application interface (WTAI).

The WAP architecture [1] was designed with a few objectives in mind: first, the limitations of the wireless communication channel needed to be addressed. These include low bandwidth by the standards of the modern Internet world, long delays, and frequently unreliable connections. For the most part, these concerns are addressed in the WAP communication protocol layers and do not usually affect WAP application developers. One exception is the low bandwidth. This is addressed by allowing WAP content, whether page layout in WML or client-side scripting in WMLScript, to be encoded into a compact binary format. This achieves a certain degree of compression over the textual form.

Compiling WMLScript [3] into bytecode form also has as an advantage in that parsing of the script does not have to be performed on the client (mobile terminals are usually limited in terms of memory and processor speed).

To ease the introduction of WAP, both for the developers providing applications and services and for the users, who must actually use them, WAP mimics to a large extent the Internet programming model. Throughout the WAP specifications, one can recognize analogies with the Internet and its technologies (see [4] for details):

› HTML vs. WML;

› JavaScript vs. WMLScript;

› HTTP vs. WSP (wireless session protocol);

› SSL/TLS vs. WTLS (wireless transaction layer security).

WAP also supports a number of content formats, such as vCard and vCalendar, that are also in use on the Internet, MIME types, and URLs.

These are only a few of the more obvious ones. This will enable developers with Internet experience to become familiar with the WAP technology relatively fast and leverage their existing experience and work.

In this chapter we will introduce components of the WAE and discuss their most important elements. In that sense this chapter can be used as a brief (though not complete) reference. We will provide background information for those who will not be involved with WAP on the actual application development level but still need a good understanding of WAP application development. We will also provide practical tips from the field that may be of use to developers that need to start developing WAP applications. The aim is to provide a brief but informative introduction.

2.2   The wireless markup language

One of the most important elements of the WAE is WML [5]. WML is described using the extensible markup language (XML) [6] as defined by the WML document type definition (DTD) that describes the format of a specific XML document. Documents that are marked up following the XML specifications are called well formed. When they also comply with their DTD, they are called valid. Some parts of WML were also based on the (now obsolete) HDML [7]. The definition of WML 1.1 deprecated many of these features in favor of more familiar constructs drawn from HTML. Developers familiar with HTML will find that WML 1.1 shares many features with HTML and should find it relatively easy to start developing with WML.

While pages on the Internet are arranged in pages (possibly compiled in a collection of frames), WML content is organized in *decks* consisting of one or more *cards*. This feature is particularly useful when WML content is viewed on devices with limited graphical capabilities and a small amount of screen area. It can also be useful in reducing the number of network transactions required to get the content into the device. Figure 2.1 graphically displays how an HTML page may be related to a WML deck of cards.

The user agent (UA) is really the browser that renders the WML content. In contrast to Web browsers, which generally provide full graphical capabilities aimed at large displays, how a UA renders WML is very much dependent on the type of device that is used. It should be clear that WML content would be rendered very differently on a text-only mobile phone than on a PDA or smart-phone type of device with graphical capabilities (supporting the wireless bitmap—WBMP) and a significantly larger screen area. This is analogous to the way Internet developers often need to think about how their content will appear on each of the popular Web browsers.

WML contains features intended to ensure that all content is readable and, more importantly, usable on all compliant devices. The distinction between readability and usability is an important one: there is little use in being able to read the text on a card if the interactive elements of the card

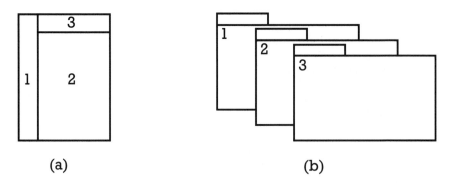

(a)   (b)

Figure 2.1   HTML page (left) compared to a WML deck of cards (right).

are unavailable on the device. To understand this, consider the use of image maps in the Web world and the problems that a text-only Web browser would have. Chapter 3 discusses the usability aspects of WAP-enabled mobile services and applications in detail.

A WML deck consists of text, with structure provided by *elements* (also called tags). Each element may have *attributes* that further specify its behavior and features. These attributes may be either required or optional (the exact requirements are given in the DTD for WML). All WML elements (tags) and attributes are case sensitive.

Attributes are often used by WML to provide hints to the browser as to how the content should be rendered. It is considered good practice to make use of these hints wherever possible, as they help to ensure that the content is rendered well across a wide range of WAP devices.

2.2.1 Decks of cards

As mentioned, the markup part of WML consists of tags (the markup) with attributes conveying additional information. All WML tags may have two standard optional attributes: `id`, which can be used to uniquely reference a particular element within the deck, and `class`, which is intended for use by the server side.

A WML file must be enclosed within `<wml>` and `</wml>` tags. Within this pair, cards are created by the `<card>` and `</card>` tags. As well as the two standard attributes (`id` and `class`), the card element can have a `title` attribute that can be used by the UA in rendering the WML content. The `id` attribute for cards can be used to reference the card directly from a URL (e.g., www.wap.net/ index.wml#welcome).

In WML it is also possible to define a template for all cards within a deck, in which certain WML event bindings can be predefined so that they need not be repeated later in every card. This is called *shadowing* in WAP parlance. These event bindings will have to be marked up between the `template` opening and closing tags. For instance, the next example specifies a "back" control for every card in the deck:

```
<template>
  <do type="prev"   name="back"   label="Back">
    <prev/>
  </do>
</template>
```

It is possible to override the WML elements specified in the template by defining an event binding for the same event. In the next example we override (in a single deck) the **do** element in the template by redefining one with an identical **name**:

```
<do type="prev"   name="back"   label="Return">
   <go href="start.wmlc"/>
</do>
```

Note how this allows us to change the task bound to the **do** element. The **do** element will still be present in this card, but its behavior will differ from the default case set in the template.

WML files can contain a **head** element that specifies information relevant to the entire deck of cards. This provides a mechanism for including various types of meta-information related to the deck, such as access control information, details of the character set, and a variety of other data.

2.2.2 User input

WML provides a set of elements for receiving input from the user. These elements are similar to those found in HTML forms, but the details of the implementation are very different: instead of merely being able to send the information to a remote server, it is stored on the user agent in "browser variables." More information on variables can be found in Section 2.2.5.

There is an **input** element supporting the entry of numeric or alphanumeric input data. Attributes on the element can indicate what type of input is to be accepted: numeric, alphanumeric, or both, and even specify the exact **format** of the user input that is to be accepted (for example, a date entered as YYYY-MM-DD), whether empty input is acceptable as well, and the maximum length of the input.

Another form of input is provided by the **select** element, which presents the user with a set of options from which to choose. For example:

```
<select name="animal"   title="Select Animal">
   <option value="w">Wombat</option>
   <option value="b">Badger</option>
</select>
```

The input as well as the **select** and **option** elements can be tailored to the exact requirements using a relatively large number of attributes.

When developing WAP applications, user input is often a major considera-tion, and knowledge of the options available for tuning the various inter-face constructs is very important, especially given the wide range of WAP devices that will become available. Intelligent use of the attributes available will maximize usability across the full range of devices.

2.2.3 Task invocation

One of the differences between WML and HTML is WML's concept of a *task*. In HTML, the range of controls available for user interaction is either hyperlinked text or images, which when activated will send the browser to a different URL or form button elements.

WML generalizes this behavior by adding the concept of a task, which can be bound to a given hyperlink or other interface widget, and can perform various actions upon the browser.

The HTML behavior, where activated hyperlinks will simply send the browser to a new URL, is provided by the **go** task in WML. When this task is executed, the browser will attempt to fetch and display the URL given by the task's **href** attribute.

The following gives an example of how a behavior similar to that of HTML forms could be achieved, using the POST functionality of the **go** task. Note that the structuring of the data in the POST is not implicit in the structure of the input fields, as it is in HTML, but must be specified explicitly with a number of **postfield** elements, each specifying a **name** and a **value**.

```
Parameter1: <input key="par1"/><br/>
Parameter2: <input key="par2"/><br/>
<do type="accept" label="Call Script">
  <go href="/cgi-bin/script.jsp" method="post">
    <postfield name="x" value="$(par1)"/>
    <postfield name="y" value="$(par2)"/>
  </go>
</do>
```

The other available WML tasks are **prev**, which sends the browser to the last card in its history stack; **refresh**, which forces the browser to redraw its display with the latest values of all the browser variables (more on variables in Section 2.2.5); and **noop**, which does nothing (no opera-tion). Calls to WMLScript are handled as a special case of the **go** task.

Tasks may be invoked at many points in the WML browser: by events (Section 2.2.4), such as when a selection is made on a **select** element, as the result of a script action, or as the result of a **timer** expiration (see Section 2.2.4), and directly by an element called **do**.

The **do** element provides a user interface element that can be activated by the user, and which invokes a task when activated. These elements may be rendered in many different ways: as graphical buttons, soft keys, or voice commands. Figure 2.2 shows how different browsers may render the same piece of WML code.

To assist the user agent in selecting the most appropriate form, the **do** element provides an attribute **type,** which offers a hint as to the function that it will perform. For example, a **do** element whose type is **accept** could be bound to the "yes" button on a mobile phone. Other types include **prev, help, reset**, etc., all of which may be used by the UA in the rendering of the **do** element. An optional **label** may be supplied as an attribute to provide a suitable text for the element.

In addition to the use of **do** elements as a means of invoking tasks, the hyperlink concept can also be used to initiate actions, using WML anchors represented by the **a** or **anchor** WML elements. In fact, the **a** element simply allows for a shortened notation of **anchor** tag. The following WML constructs all have exactly the same functionality, each displaying the text "hyperlink" as a hyperlink, which will send the browser to the card with the name "card2" when activated.

```
<anchor>hyperlink<go href="#card2"/></anchor>
<anchor><go href="#card2"/>hyperlink</anchor>
<a href="#card2">hyperlink</a>
```

2.2.4 Events

Tasks can be invoked by certain events that represent state transitions within the UA. This is specified in WML using the **onevent** element. A number of event types are defined: **onpick** (an **option** element is selected by the user), **onenterforward** (a card is entered via a **go** task or through direct user intervention), **onenterbackward** (a card is entered using the **prev** task), and **ontimer** (a timer defined with the **timer** element expires). Typical usage of the **onevent** tag may be as follows:

```
<onevent type="onpick">
  <go href="choice.wml"/>
</onevent>
```

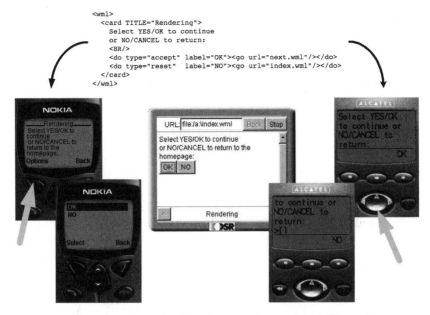

```
<wml>
  <card TITLE="Rendering">
    Select YES/OK to continue
    or NO/CANCEL to return:
    <BR/>
    <do type="accept" label="OK"><go url="next.wml"/></do>
    <do type="reset"  label="NO"><go url="index.wml"/></do>
  </card>
</wml>
```

Figure 2.2 Examples of how WML markup may be rendered differently on different model WAP phones and browsers. Images are based on phone emulators rather than actual phones. The Nokia phone recognizes the presence of two go tags and creates a list of options that can be selected. The WAR (wireless application reader) browser renders them as two buttons in the absence of a mapping mechanism to keys on a device while the Alcatel phone emulator maps both do elements to the same key. Selection is done using the navigation key in the center.

2.2.5 Variables and context

A particularly interesting and useful feature of WML is that the developer can define variables with the WML itself. All variables defined in a UA session are called the *context* of the UA. Furthermore, the value of these variables can be substituted into the WML content before it is rendered: text can be substituted into the body of a card, into the URL of an image file, or even into the URL of a go task. This is done by placing the variable in the WML code as either $varname or $(varname). If a real $ symbol is required in the text of a page, it can be represented by $$.

Values may be assigned to variables, using the setvar element within some task element. Variable names are case sensitive. For example, to set the variable location to contain the value checkout, the WML construct <setvar name="location" value="checkout"/> would be used. This can be done automatically when a card is entered by means of the `onenterforward` event, using a `refresh` task as follows:

```
<card name="card1" title="Checkout">
  <onevent type="onenterforward">
    <refresh>
      <setvar name="location" value="checkout"/>
    </refresh>
  </onevent>
</card>
```

It is also possible to clear the context of the UA (the collection of all variables and also the history stack), by setting the optional `newcontext` attribute of the `card` element to `"true"`:

```
<card title="In the beginning..."
  newcontext="true">
  ...
</card>
```

WML variables may be used for storing short-lived context during a user's interaction with an on-line application. However, since the context may be cleared by any new application, any data that are required to be persistent must be in a persistent manner, something that is currently not provided for in the WAP specifications. Such data could, for instance, be stored on the server by means of a POST operation.

2.2.6 Other features of WML

WML also has a number of elements that can be used to influence the actual layout and rendering of content. Text can be marked up by the `big`, `small`, `em` (emphasis), `strong` (strong emphasis), `b` (bold), `i` (italics), and `u` (underlined) elements. Line breaks can be forced using the `br` tag.

The paragraph tag `p` delimits a paragraph of text in the body of the document. Through a number of its attributes you can set a number of layout options for that paragraph, such as whether the text in the paragraph should be wrapped if the lines will not fit within the width of the screen, and how they should be aligned on screen.

WML also has tags to define tables (through use of the `table`, `tr`, and `td` elements) that are very similar to HTML tables. The WML ele-

ments **td** (table data) and **tr** (table row) have been introduced in the
WML 1.1 specification and they facilitate substantially easier conversion
of HTML tables to WML than was possible before. In general, the WAP
Forum has expressed the intention to work towards a convergence of
WML and HTML towards XHTML.

Special characters, such as angle brackets, nonbreaking space, and
the like, are encoded in WML in a special way similar to HTML (for
example, both ** ** and ** ** represent a nonbreaking space). Comments may be included in WML decks between **<!--** and **-->**.

The use of images is also possible in WML. WAP images are in a format called WBMP and one bit deep (black and white). Images can be
included by using the **img** tag, and a number of attributes can be set,
including the **src** attribute. Images can be aligned and their height and
width specified. The **alt** attribute is required and is used by the user
agent to display a textual alternative to the image, if image display is not
supported by the user interface. Images can also be embedded within an
anchor or **a** element, thus making them useful for navigation purposes,
although not all WAP browsers may support this feature.

To summarize, WML is a simplified yet powerful markup language
that shares many similarities with HTML, making it easy to learn for
those that are already familiar with the Web. It also provides a variety
of interesting additional features that make it more powerful in terms
of user interaction and client-side processing. The use of variables in
WML itself can make the maintenance of state-of-user sessions much
easier than using HTML, especially since these variables can be accessed
from within WMLScript programs as well. Another valuable addition is
the concept of tasks and events in WML. These features make it possible
to develop flexible applications in WML without the use of server-side
logic.

2.3 Wireless markup language script

While WML is itself a feature-rich markup language, additional application logic and intelligence can be added using WMLScript [3]. WMLScript
is a scripting language based on JavaScript [8] (or ECMAScript [9]). However, unlike JavaScript, WMLScript code is not embedded within the
WML deck, but instead it is compiled into compact binary files, called
compilation units. Transmitting these compiled units to the user agent
rather than the textual source form results in a reduction in size and

also the amount of parsing/processing required on the WAP enabled device.

2.3.1 WMLScript variables and datatypes

WMLScript is an example of a dynamically typed language. This means that while all data within the language have a type internally, variables are not explicitly typed: any variable could contain a value of any type, and a function can return a value of any type. Four main datatypes are supported: integers (32 bits signed), booleans (either true or false), floating point numbers (IEEE single precision), and strings (of arbitrary length subject to memory restrictions on the user agent).

There is also a pseudo-datatype, *invalid,* which can result from an illegal operation (such as division by zero), or a failed attempt to coerce a datum from one type to another. Normally, the script interpreter will implicitly attempt to convert the values into the type required. If this cannot be done for some reason, then the result is invalid. An attempt to use an invalid value in an operation results in invalid for the result of the operation (for example, the expression 1/0+2 will be invalid since 1/0 is invalid).

Variables must always be declared in WMLScript, though they do not necessarily have to be initialized to a specified value (if no initialization is provided, an empty string is assumed). The scope of a variable runs from the location in which it is defined until the end of the function in which it is defined. There is no concept of a global variable, though this can be circumvented by storing WMLScript variables in WML variables that make up the browser context.

WMLScript does not support arrays natively, but provides flexible string manipulation and parsing capabilities to remedy this situation. Arrays can be easily simulated by creating a string containing the elements separated by some symbol. For instance, the array of real numbers (which could be the prices of various products) containing 12.95, 8.89, and 589.0 could be represented as the string "12.95/8.89/589.00". This array can then be accessed and manipulated using standard library functions to insert, append, remove, and replace elements.

WMLScript supports the normal arithmetic, bit-wise, comparison, logical, and conditional operators provided by JavaScript, and also two special operators related to the dynamic typing scheme. The typeof operator may be applied to any value and returns an integer in the range zero to four corresponding to the current type of the value. The boolean isvalid operator may also be applied to any value, and returns false if the value is invalid and true otherwise.

2.3.2 WMLScript functions

WMLScript compilation units consist of one or more WMLScript functions, whose names must be unique within a single unit. Function declarations cannot be nested, parameters are always passed by value, and all variables and parameters are local to that function. WMLScript functions always return a value: if none is explicitly specified, then the empty string is returned.

To call a function from a different unit, or from WML, the declaration of the function must be prefixed with the extern keyword. If this is not done, the function's name is not stored in the bytecode file, resulting in a saving of space, but with the penalty that the function is not accessible from outside that unit.

When calling a function in a different compilation unit, the pragma use url is used to specify the URL of the external compilation unit. For example:

```
use url verify http://www.wap.net/libs/sec/
verify.wmlsc
function process (params)
{
    var ver=verify#test(params);
    ...
}
```

WMLScript also provides a set of standard libraries (see Section 2.3.5 and [9]) that are called as follows: LibName.functionName (params).

2.3.3 Differences between ECMAScript and WMLScript

ECMAScript is a standardized scripting language and a superset of Java-Script. In this section differences between WMLScript and ECMAScript are briefly summarized in Table 2.1.

2.3.4 WMLScript statements

WMLScript provides a number of statements that should be familiar to anyone with knowledge of JavaScript:

Table 2.1
A Summary of the Differences Between ECMAScript and WMLScript

WMLScript	ECMAScript
Statements always followed by semicolon	Semicolons optional in some cases
Comments between /* and */	HTML comments also supported
Local variables only	Local and global variables supported
Compulsory declaration of variables	Implicit declaration supported
Integer and floating point number types	All numbers internally 64-bit floating point
No native support for arrays	Support for arrays
The number of arguments accepted by a function is fixed	Supports variable number of arguments in function call
Pragma support	No pragma support
$ not allowed as part of identifier	$ allowed in identifier name at any position
The type of operator returns an integer representing the type of the argument	The type of operator returns a string representing the type of the argument
Functions in WMLScript are not datatypes	Functions are a special datatype
Number of arguments in function call must match that in function definition	Number of arguments in function call may differ from that in function definition
No support for objects	Support for objects
Does support the invalid type	Does not support the invalid type
Supports integer division (div)	Does not support integer division

- Empty statement: a single semicolon (;);
- Expression statements: any expression can be used in place of a statement;
- `block` statement: a number of statements enclosed in curly brackets `{` and `}`;
- `if ... else`;
- `for`;
- `while`;
- `break`;
- `continue`;
- `return`.

There is also the `var` statement used to declare and optionally initial-ize variables:

```
var foo, bar=3;
var wibble, wobble, wubble;
```

2.3.5 Wireless markup language script standard libraries

WMLScript programs can make use of a set of libraries that contain addi-tional functions to those present in the WMLScript core definition. The following WMLScript standard libraries are defined [10]:

- *Lang:* core functions such as random number generation and func-tion exiting;

- *Float:* floating point mathematical operations;

- *String:* various string operations—also primitives useful for array operations;

- *URL:* functions for parsing URLs, escaping and unescaping, and downloading content directly into WMLScript variables;

- *WMLBrowser:* interaction with the user agent—refreshing, getting and setting variables, changing the displayed URL;

- *Dialogs:* simple dialogs to alert or prompt the user.

As the WAP specifications evolve, it is likely that other WMLScript libraries will be defined. An example may be the WMLScript crypto library that gives the WAP developer access to cryptographic functions and provides security functionality on the application level in WAP 1.2.

2.4   Byte-encoded WML and compiled WMLScript

One of the objectives of WAP is to provide a service creation technology that is optimized for use over wireless communication networks: net-works that currently have a limited bandwidth available. While optimiza-tion can take place on the communication protocol level, it can also be accomplished by compressing the content to be transported.

The latter optimization is defined in the WAE and is achieved in two ways: WML content is encoded into a compact binary format, and

WMLScript is compiled into a special bytecode that can be run within the WMLScript virtual machine on the WAP-enabled device.

2.4.1   Encoding of WML

The byte-encoded format of WML is based upon the WAP binary XML content format (WBXML) [11]. WML is encoded by replacing parts of the WML (tags, attribute names, and attribute values) by single-byte tokens. For instance, the WML encoding process replaces the WML attribute and value `type="onenterbackward"` with the single byte 61. In this example, 21 bytes have been saved over sending the WML directly in ASCII format. Plaintext in WML code can be compressed using string tables. In general, the size reduction will be less spectacular, especially when the WML does not contain a lot of text information, since this process has little effect on the text in the body of the document.

The WAP specifications define that user agents must accept tokenized WML as well as plaintext WML. It is likely that in practice only encoded WML (encoded on the fly by the WAP gateway, for instance; see Chapter 5) will be transferred to the UA since this allows a significant reduction in the amount of data transferred, and hence transfer time and cost incurred by the user.

2.4.2   Compilation of WMLScript

WMLScript content must also be compiled into binary format before being transmitted to the user agent. A virtual machine (VM) interpreter in the user agent then interprets this bytecode in order to execute the program. The situation is not unlike the way Java code is compiled before it is interpreted by the Java VM.

Unlike the encoded form of WML, the WMLScript binary format is not simply a compact representation of the original textual data, but is instead a set of encoded instructions for a virtual machine. Several WMLScript compilers have the ability to output a textual representation of these instructions, as seen in Figure 2.3.

2.5   Overview of the wireless telephony application interface

WTAI (see Chapter 4) provides WAP developers with the means to access the telephony functions of the device running the WAP application. For instance, an application might make a voice call on behalf of the user,

```
/*  calculate n factorial:          */
/*  1*2*3* ... *n                    */
extern function factorial  (n)
{
if (n<0)
return invalid;

var total = 1;
while (n>1)   {
total *= n;
--n;
}
return total;
}
load_var  0#  load n
const_0#  constant 0
lt
tjump_fw label1#  jump unless n<0
const_invalid#  constant invalid
return#  return invalid
label1:
const_1#  constant 1
store_var 1#  store into total
load_var 0#  load n
const_1#  constant 1
gt
tjump_fw label2#  jump unless n>1
label3:
load_var  1#  load total
load_var  0#  load n
mul#  multiply them
store_var  1#  store into total
decr_var  0#  decrement n
jump_bw  label3#  jump always
label2:
load_var  1#  load total
return#  return total
```

Figure 2.3 An example of WMLScript source code and a representation of the instructions generated by the compiler, together with comments on the instructions.

send a text message, or access the phonebook. The WTAI is part of a larger set of specifications, which allow for a truly seamless integration of the WAP browser and the normal functions of the phone in a way that was never possible with HTML and the Web.

WTAI functions may be called from WMLScript as standard library function calls, or they may also be initiated by accessing specially formatted URLs from within WML content.

2.5.1 WTAI libraries

The WTAI functions are collected in WTAI libraries, which are in turn divided into three different types: public, network common, and network specific. Public functions are universally available high-level operations available to all WAP applications. Network-common and network-specific functions are at a lower level. In the case of network-specific functions, these functions are specific to the type of wireless network at hand (e.g., GSM). The latter two are not of particular interest to developers of general applications; they provide in-depth access to the internal workings of terminals, but are inaccessible to most applications. Table 2.2 gives an overview of the available WTAI function libraries.

2.5.1.1 Public WTAI

As an example of a WTAI possibility, we demonstrate how WTAI functionality can be accessed using methods from the public WTAI function library using the make call function, which initiates a voice call to a specified phone number. The user must explicitly acknowledge the operation.

This function may be accessed from the WMLScript as the following library call:

```
WTAPublic.makeCall("5551234");
```

Alternatively, it may be accessed from WML through the following URL:

```
wtai://wp/mc;5551234;
```

Table 2.2
WTAI Function Libraries

Function Library	Name	Description
Public WTA	wp	Publicly available WTAI functions
Voice Call Control	vc	Call setup and control during connection
Network Text	nt	Sending and receiving text messages
Phonebook	pb	Phonebook management
Call Logs	cl	Access to device call logging data
Miscellaneous	ms	Miscellaneous

In conjunction with the ability of the public WTAI to send DTMF signals through an open voice channel, this could allow a WAP application to directly interact with a legacy DTMF-based help system to provide a context-sensitive help function to the user.

2.6 Migrating from WWW to WAP

In many cases WAP will be attractive to parties already offering content and services over the Internet and who wish to exploit the opportunities of the new mobile channels. For developers, this could mean the need to think about migrating from the Web, where services are created with HTML and JavaScript, to a WAP-based model, where WML and WMLScript are the available technologies. Arguably, such developers will want to create WAP services that are very similar in terms of usability to the existing Web-based ones.

Thus, it may be of interest to briefly discuss the differences between both HTML and WML, and between JavaScript and WMLScript. The deployment of existing CGI (common gateway interface) applications in WAP is also outlined here.

2.6.1 WML and HTML

WML and HTML, both being markup languages with hypertext elements, have a lot in common. Although WML lacks the same range of possibilities for controlling the rendering of content, it has gained useful functionality in the form of variables and tasks, allowing considerably more client-side processing to be done on the data.

Conversion (or filtering) of HTML to WML is by no means a straightforward operation, though the WAP Forum works towards a convergence between HTML and WML. In the restricted case where the HTML formatting of information (e.g., by means of templates) is such as in a corporate intranet environment or Web site generated directly from database content, conversion may be considerably simplified. However, it is unlikely that filtering will be a generic solution that can be applied to the wide variety of Internet sites present today. This means that new filtering rules need to be specified for every new individual HTML source. Nevertheless, HTML to WML conversion may be a viable solution in the cases where the use of difficult HTML constructs is avoided, and when producing a native WML solution would be too labor-intensive.

A second issue arises from the fact that WAP browsers in general will have significantly less screen real estate compared to most Web browsers. Besides conversion of the markup, the actual content of the page will need to be split up into different WML cards. It is not necessarily clear how and where these divisions should be placed.

With the conversion of the markup and the division of the actual content or application logic into chunks that make sense to a WAP user, the question arises as to how the user will experience the new WAP application. An on-line application that works well when viewed in a Web browser on a PC may be unusable on a WAP-based mobile phone. Rather than a simple split of the HTML page into a number of WML cards, the user of the WAP application needs to be offered a simple path through the content or application. In some cases this may be at the cost of a more frequent client-server interaction and this may cause additional delays compared to the original Web-based version of the application. In the worst-case scenario, these considerations may even call for significant design changes to the way the service is provided.

2.6.2 WMLScript and JavaScript

Where the conversion from HTML to WML may be a realistic proposition, the automatic conversion of JavaScript to WMLScript is not viable. As a simple case, all JavaScript content can be stripped from the markup, but certain operations will need replacing with WMLScript, for example, when the JavaScript is used to check parameters for validity before they are sent to the server. JavaScript is usually embedded within the HTML markup. Any attempt at automatic filtering will need to replace a call to a JavaScript function to a call to some prewritten WMLScript.

2.6.3 Using CGI and WAP gateways

Currently, there are at least two ways in which WAP services can be deployed. In one scenario, there is the situation where applications based on conventional Web servers are configured such that they provide content in WML and WMLScript format rather than HTML and JavaScript. Using this model, a WAP gateway provides protocol conversion services (from HTTP to WSP) and possibly converts content from HTML into WML. This WAP gateway will also encode the WML in its bytecode format and may need to compile the WMLScript before sending it to the user agent. The second scenario consists of a dedicated WAP server that is able to host applications for itself and provides an API to those applications to allow them to access features of the server.

In the first case it is likely that the WAP gateways will be installed within the network of the wireless operator, or at some other access or dial-up point, such as a traditional Internet service provider (ISP). The gateway operators can then make access points available for the service providers, each of which will have its applications running on a remote site and accessed through the Internet, a leased line or a virtual private network (VPN).

In the second case service providers would either host the WAP server themselves or would rent space on another party's server in a way similar to how third-party Web hosting functions in the Internet world. In that scenario the service provider may also have to provide dial-up access to the users of its services.

The use of a WAP gateway as a protocol and content converter provides a relatively easy migration path for those providers who already have a service offering over the Internet. It is merely necessary to change the format of the content that they provide, depending on the browser that accesses their URL. This is particularly easy when content is published dynamically from databases or available at the source in some abstract form, such as XML. There are many more issues surrounding offering WAP services, such as user provisioning and billing, and we refer to Chapter 5, where the role and functionality of WAP gateways is described, albeit more from the network operator's point of view.

2.7 Markup languages and XML

We have briefly outlined the possibilities in migrating Web applications to WAP and suggested that one of the reasons that WAP has enjoyed such great interest, especially from the providers of content and services and from the developer communities, may be the relative ease of migration. However, the multitude of available channels for providers to deliver their applications and services still raises many issues related to how these channels should be dealt with, while simultaneously minimizing development and maintenance costs and maintaining a high degree of flexibility with respect to future technological developments.

For many, the answer lies in taking a step back from the presentation of their content and using a more generic representation of their content at the back end of their systems. XML [6] provides just such an abstract representation. XML allows the presentation of data to be separated from the data themselves, and gives XML authors and developers the freedom to effectively structure their information independently of how it is to be displayed.

A popular application of XML is to store certain information in a format portable across platforms, applications, and display methods. For instance, an example of XML to store information about traffic delays on a highway could look like:

```
<TRAFFICINFO>
  <ROAD ID="M4" START="Chiswick" END="Heath-
row"/>
  <DELAY TIME="80" UNIT="minute"/>
  <CAUSE>
    Poor visibility
  </CAUSE>
  <CONGESTION LENGTH="2" UNITS="mile"/>
</TRAFFICINFO>
```

2.8 User agent capabilities and content negotiation

One thing that a WAP developer must consider over and above what a Web developer must consider is that WAP devices will vary enormously in their capabilities, from different screen resolutions, through different methods of user interaction, to different sets of optional features that are implemented. The production of optimal WAP content requires knowing the characteristics of the target device.

This can be achieved in one of two ways. First, a user agent will identify itself by means of protocol headers while communicating with the gateway (e.g., the HTTP/1.1 "User_Agent" header), and this could be compared with a static table of device characteristics to determine the capabilities of a given user agent. The exact WML formatting can then be determined dynamically. This method has been widely used in the Internet world but requires a regular updating of the table of characteristics to keep up with the latest browsers and is impractical as a result.

The second way that WAP makes the exchange of device characteristics possible is that the initiation of a connection between the client and the gateway includes the negotiation of a set of mutually acceptable capabilities. The set of capabilities includes the versions of the various content formats and protocols that are supported, as well as whether such features as images and server push are implemented. This mechanism is defined in the WAP specifications [12].

The exchange between WAP client and server of information regarding the capabilities of the client and server is described in the WSP [13] and proposed UAProf (user agent profile) [12] specifications, but this information may often not be of particular interest to application developers, other than the list of negotiable capabilities.

The profile of a user agent may also be stored in a special format and referenced via a URL. This allows for more efficient negotiation over the air: rather than having to send all of its characteristics over the slow link, a user agent need only send a URL pointing to the set of capabilities for that model of device or that version of the browser software. These do not change from one session to the other.

2.9   Miscellaneous elements of WAP of interest to developers

2.9.1   Push

WAP provides the ability for a server to send content to the user agent asynchronously and without an explicit request. For example, this could be used to signal the arrival of e-mail, a financial transaction with certain characteristics, or even notable events in a sports match.

This functionality is partly incorporated into the WSP [13] and push and is included in the capability negotiation described in Section 2.8. Chapter 6 will describe in greater detail push services implemented in WAP.

2.9.2   Wireless session protocol and HTTP headers

WSP [13] contains certain elements that may be of direct interest to WAP application developers. The WSP is a communications protocol that is based on, and in many ways similar to, the Internet standard HTTP/1.1 [14] with some additions and modifications to optimize its use over wireless channels. These modifications deal with the well-known limitations of wireless communication channels such as low bandwidth and susceptibility to loss of connection due to poor coverage or cell overloading.

WAP uses what can be called a split-server model of content distribution, where the content resides on an origin server or application server and is transferred via HTTP or secure HTTP to the WAP gateway for forwarding to the client.

Information between two resources can be exchanges in HTTP and WSP packet headers. Developers can manipulate the use of these to enhance their WAP applications. The accept header gives the range of MIME media types supported by the sending party; accept-charset gives the range of character sets supported by the sending party; and accept-language lists the sending party's preferred languages in order.

The WSP specifications state that certain HTTP headers (such as those for cache control mentioned in Section 3.9.4) will be transformed to the appropriate WSP headers when passing through a compliant gateway. This allows developers to use standard Web servers as WAP origin servers with a little reconfiguration. The fact that one can define and implement new HTTP headers in version 1.1 means that a CGI script used for generating dynamic content can simply add extra HTTP headers to the top of the request, and these will get to the user agent intact. Similarly, a gateway may request a WML page (on behalf of the WAP UA) and may provide information in additional HTTP/1.1 headers such as the MSISDN or unique user ID of the user. The application, in return, may provide billing information, such as the price of the requested item, in extra HTTP/1.1 headers.

2.9.3 Binary encoding of wireless session protocol headers

The WSP uses headers attached to requests and responses to communicate additional information between the client and gateway. Optimization with respect to the low bandwidth generally available over wireless channels is achieved by a binary encoding of these headers, in a similar way to that used for WML. This results in a dramatic reduction in the amount of data that must be sent back and forth between server and client.

2.9.4 Cache control using wireless session protocol headers

WSP inherits from HTTP two headers used to control the behavior of caches along the route from origin server to user agent. These two headers, expires and cache control, have identical semantics to their HTTP counterparts and should be used by application developers wherever necessary to maintain the freshness or confidentiality of data. For example, dynamic information such as traffic information, stock quotes, or sports scores either should not be cached or should expire quickly. Slowly changing information such as cinema or television listings should expire more slowly (after a day or more), but information such as a film review

or a list of attractions for a major city can be expected to change suffi-
ciently slowly that caching would not be a problem.

2.10   Available software tools

A number of software tools are available for WAP developers: some
of these tools have been included on the CD-ROM accompanying this
book. These tools vary in the way they look and what they aim to
achieve.

Some attempt to implement a comprehensive set of tools for developing applications (including support for SIM toolkit, for instance), while
others may feature browsers that do not even support WMLScript. Many
tools also aim to provide a very graphical user interface, including phone
simulators and editors for modeling new phones. Other tools focus on
delivering the best performance in the WML encoder and the WMLScript
compiler to achieve a significant reduction in the size of the binary
content.

While it may be useful to get an impression of how their applications
may look on different phones, developers should be cautious not to
develop applications for a particular type of device. To do so would be to
ignore the true potential of WAP. This chapter contains information on
how to develop applications for a number of different types of mobile
WAP devices.

For a WAP developer, it may be a good idea to use more than one
WAP software package. Experience has shown that some packages
may contain bugs and vary in their implementation of certain WAP
features. Trying WAP applications on a number of WAP software tools
and evaluating their behavior may be the way to avoid unpleasant
surprises later. As WAP is still an evolving standard, so are the developer tools. As a result, it is probably useful that developers try out as
many of the kits as possible and do not use a single package for
everything.

2.11   WML language reference

In Table 2.3 we provide a very brief summary of all WML elements in the
WML 1.1 specifications. Many elements have associated subtleties that
require the developer to consult the actual specifications for optimal
results.

Table 2.3

Summary of WML Elements in WML 1.1 Specifications

Tag Name	Attributes	Default	Description
\<a\>	href		URL to be navigated to upon selection of the anchored link.
	title		Title of the anchored link.
\<access\>	domain		Domain from which access is allowed.
	path		Path from which access is allowed.
\<anchor\>	title		Title of anchored link.
\<b\>			Renders fonts bold-faced.
\<big\>			Renders fonts big.
\<br\>			Line break.
\<card\>	title		Title of card.
	newcontext	false	Specified whether browser context should be reset.
	ordered	true	Hint for rendering of content by browser.
	onenterforward		URL to be navigated to when card is entered.
	onenterbackward		URL to be navigated to when card is left.
	ontimer		URL to be navigated to when time expires.
\<do\>	type		Type of task. Can be 'accept', 'prev', 'help', 'reset', 'options146', 'delete', or 'unknown'.
	label		Label to be used for rendering purposes.
	name	'type'	Name of 'do' event binding.
	optional	false	Indicates whether this element is optional for rendering by browser.
\<em\>			
\<fieldset\>	title		Title for this WML element.
\<go\>	href		URL to be navigated to when invoked.
	sendreferer	false	Whether or not the referring URL is to be sent when invoked.
	method	get	HTTP submission method. Can be 'get' or 'post'.
	accept-charset	unknown	Character encoding that must be accepted.
\<head\>			Contains (meta) information about the deck.
\<i\>			Renders font in italic.
\<img\>	alt		String for alternate rendering when image cannot be displayed.
	src		Image URL.
	localsrc		Internal representation of image.
	vspace		White space to be inserted above and below image.
	hspace		White space to be inserted left and right of image.
	align	bottom	Image alignment.
	height		Height of image.
	width		Width of image.

Table 2.3 (continued)

Tag Name	Attributes	Default	Description
<input>	name		Name of the variable to hold input.
	type	text	Type of input. Can be 'text' or 'password'.
	value		Default value of variable.
	format		Format of input.
	emptyok	false	Specified whether empty input is acceptable. Can be 'true' or 'false'.
	size		Width of input area.
	maxlength		Maximal number of characters as input.
	tabindex		Tabbing position of this element.
	title		Title of 'input' element.
<meta>	http-equiv		To be read as HTTP header.
	name		Name of 'meta' tag.
	forua		Specifies whether this meta-information is intended for the browser. Can be 'true' or 'false'.
	content		Value of meta-information.
	scheme		Scheme to be used.
<noop>			No operation task.
<onevent>			Task binder.
<optgroup>	title		Title of this element.
<option>	value		Value to be stored in variable when this option is selected.
	title		Title of this option.
	onpick		URL to be navigated to when this option is selected.
<p>	align	left	Text alignment mode for the paragraph. Can be 'left', 'center', or 'right'.
	mode	wrap	Line-wrap mode for paragraph. Can be 'wrap' or 'nowrap'.
<postfield>	name		Name of the data to be posted.
	value		Value of the data to be posted.
<prev>			
<refresh>			Render the page again with the current value of the variables.
<select>	title		Title of this element.
	name		Name of the variable to be set as the result of a selection.
	value		Default value of the variable to be set.
	iname		Name of the variable to be set with the index of the selected option.
	ivalue		Default value of the variable to be set with the index of the selected option.
<refresh>	multiple	false	Multiple selections. Can be 'true' or 'false'.
	tabindex		Tabbing position of this element.
<setvar>	name		Name of variable that is set.
	value		Value of variable that is set.
<small>			Renders font small.

Table 2.3 (continued)

Tag Name	Attributes	Default	Description
<strong>			Renders font emphasized.
<table>	title		Title of table.
	align	left	Alignment within each column.
	columns		Number of columns in table.
<td>			Table data.
<template>	onenterforward		URL to be navigated to when card is entered.
	onenterbackward		URL to be navigated to when card is left.
	ontimer		URL to be navigated to when time expires.
<timer>	name		Name of the variable holding the timer value.
	value		Value of the timer before expiring.
<tr>			Table row.
<u>			Renders font underlined.
<wml>			Start of WML code.

References

[1] Wireless Application Protocol Architecture Specification, WAP Forum, Version 30, April 1999.

[2] Wireless Application Environment Overview, WAP Forum, Version 16, June 1999.

[3] Wireless Markup Language Script Specification, WAP Forum, Version 17, June 1999.

[4] Wireless Application Environment Specification, WAP Forum, Version 25, May 1999.

[5] Wireless Markup Language Specification, WAP Forum, Version 16, June 1999.

[6] Bray, T., et al., *Extensible Markup Language (XML)*, W3C Proposed Recommendation 10, February 1998, REC-xml-19980210, February 10, 1998.

[7] King, P., et al., *Handheld Device Markup Language Specification*, April 11, 1997.

[8] Flanagan, D., *JavaScript: The Definitive Guide*, Sebastopol, CA: O'Reilly & Associates, 1997.

[9] Standard ECMA-262: ECMAScript Language Specification, ECMA, June 1997.

[10] WMLScript Standard Libraries Specification, WAP Forum, Version 17, June 1999.

[11]   Binary XML Content Format Specification, WAP Forum, Version 16, June 1999.

[12]   User Agent Profile Specification, WAP Forum, Version 10, November 1999.

[13]   Wireless Session Protocol Specification, WAP Forum, Version 28, May 1999.

[14]   Fielding, R., et al., *Hypertext Transfer Protocol 1.1—HTTP/1.1*, January 1997.

Contents

Designing Effective User Interfaces for WAP Services

Marcus Taylor, Ian Hosking, and David Brazier

"Three clicks and you're out."

(After California Penal Code Section 667, known as "Three strikes and you're out.")

3.1 Introduction

Presented with the WML specification (see Chapter 2 for details), one might imagine that there is no scope for user interface design of WAP services. Immediate reactions might be:

› Phones have such small screens; there's no room for any fancy graphics.

› The specification gives the device such flexibility for presentation detail that there's no scope for design at the application level.

› All WAP services will have simple menu-driven interfaces.

› We can just port our Web interface onto WAP.

Indeed, experience gained to date in developing WAP services suggests that these reactions are common. To be fair, of course, most WAP services may not have been very developed in terms of usability.

First, we should reflect on the immediate reactions mentioned previously.

1. *Phones have such small screens that there's no room for any fancy graphics.* The small screens on phones do preclude large graphic images. Also, they tend to be monochrome, so color images will not be rendered properly. Finally, typical network bandwidth dictates small images used sparingly. However, the above statement implies that user interface design is mainly about fancy graphics, which it is not. User interface design is about helping the user carry out a task. The small screens make the design task a particularly demanding one. So a better reaction would be: Phones have such small screens, how will the application be able to use the space effectively?

2. *The specification gives the device such flexibility for presentation detail that there's no scope for design at the application level.* The WAP specification does allow for a great deal of variation in presentation styles on the device, for very good reasons. The specification is intended to be capable of implementation by a range of devices. Even among one class of device (mobile phones, say, with similar physical characteristics, for example, screen size) the manufacturers' own user interface styles must be carried through to the WAP service. These styles have been developed by the manufacturers to make the phone's own facilities easy to use and are thus a key differentiator in the marketplace. As our immediate reaction implies, we can't fight against this: we would be trying to pre-empt the device's user interface styles and risk confusing the user. However, we are in danger of falling into a trap similar to some developers at the advent of graphical user interface (GUI) systems such as Microsoft Windows, which is assuming that the new GUI and all its standard components—buttons, list boxes, and so on—do all the user

interface design for us. Unfortunately, the handy components (in WAP as in Windows) allow us to put together a bad interface much more easily than design a good one. Our reaction should be: The specification gives the device such flexibility for presentation detail that we will have to work hard to design a good interface.

3. *All WAP services will have simple menu-driven interfaces.* Simple menu-driven interfaces are great for some tasks and obviously sit well with the facilities on most phones. Good examples, where the user can simply navigate a hierarchy of information—such as cards in a contact list—are easy to find. However, not all tasks fit this model. Games are a straightforward counterexample: if the target I want to shoot is to the left, I don't want to press Options and find Move Left on a menu, I want to press the left button. More businesslike examples include traffic information systems. Our reaction should be: Simple menu-driven interfaces are often a good solution, but always look for alternatives.

4. *We can just port our Web interface onto WAP.* Since WAP is often portrayed as synonymous with mobile Internet, this seems to be a straightforward approach. After all, WML is based on HTML, isn't it? In spite of this, the Web interface probably addresses a different set of tasks than those required by the mobile user. If there is a Web page for setting up a new account—with name, address, three phone numbers, mother's maiden name, and so on—would we really expect that to be used by someone on a phone? Probably not. So we should reconsider the tasks the user wants to perform when mobile and design around those. Of course, we hope that the back-end processing built for the Web pages will be useful for the mobile service, but we might have some work to do there, too. Thus, an existing Web interface might provide some input to the WAP interface design.

Now that we have cleared the air of these unhelpful reactions, we can think about how to go about designing our WAP interface. First, we consider the process we need to adopt to get the right people working together to produce the best service. Then we can look at general design principles and some more detailed design considerations: navigation, presentation, and input.

3.2   The user interface design process

User interface design is about solving a mixture of technical and business problems, which can only be solved in the right environment. A logical way to go about creating the ideal environment is to involve the entire organization in a holistic approach (i.e., involving both marketing and technical departments in an iterative design and implementation methodology as opposed to the "waterfall model" of systems design). Each department has a role to play in providing ideas, potential requirements, and expertise.

3.2.1   Holistic process

Sales and marketing are typically the people that get the budget for developing an idea, while other departments will have specific requirements for the service. What these respective departments or personnel need are tools and guidelines that enable them to steer those requirements into a product which has a high level of usability and ease of interaction.

There is no real magic wand solution to creating a good user interface. Once you go through the necessary processes and use the required tools to cover all the angles, you can begin to prototype on the basis that the system will not work the first time. By using the process of iteration and getting both the relevant departments and potential users involved at every stage to offer input and their point of view, be it positive or critical, the user interface will evolve from an initial concept to an efficient working system.

The holistic approach to user interface design is based on common-sense principles. You can break down the task into the components shown in Table 3.1.

Table 3.1
Design Process Components

Gather customer requirements	This can be derived from market research or existing user profiles of a segment of your customer base.
Completing the rough cut, high-level design	This can be achieved using paper and pen to sketch out the essential tasks.
Designing the details	This is where technical considerations for viable requirements gathered are turned into an end product.

There are a lot of different tools that can be used with a holistic approach. In terms of creating the requirements, if you already have an existing product, you will have a lot of background information that you can use. Additionally, look to competing products to compare and contrast. If you have a customer support center or help desk, they will have gathered or classified different types of problems that customers may have experienced. User groups and focus groups can provide an incredible amount of information by putting people together in a room and asking them to brainstorm. There is a danger in asking users what they want, though. Potential users of a system will generate a list of demands, but when they have to use the system, they realize that what they want is often not what they need.

The other point about using a holistic approach is that it is multidisciplinary. You need a software developer, marketing and sales people, a technical author, and a graphic designer. All of these people bring different perspectives and skill sets that have an important role to play in developing a good quality service or product. If you can get all of those skill sets together in one room, then you have a very good chance of bringing out a very good product.

Promoting a dialogue between departments is also vital to ensure that everybody is speaking the same language, and this communication, be it by e-mail, memo, or face-to-face, will help to achieve the common goal. The end result will be a user interface that not only interacts with the system, but also is an interface where a developer can speak on the same terms with a marketing person. It thus creates a forum where people are speaking in a constructive way to make a good product or service.

3.2.2 Customer satisfaction

When designing a user interface for a particular service, the development team should always remember that one day this service will have to be easy to use and able to be used on a daily basis without expert knowledge. Through the ease of use and ease of learning, one can encourage further use and adoption of enhancements of that product or service later on.

The real test of any product or service is when your customers have to pay for it. While a prominent feature of the Internet is the concept that the information provided by a Web site is free, it is inevitable that most WAP services will have a commercial nature. Another factor to consider is that connection time for mobile users may be at a premium, compared to that of an Internet connection on a fixed-line phone. Even where cost is not an issue, the users' time probably will be—one reason they are using the WAP service is that they don't have time to use the Internet.

Once paid for, does the customer feel there is a good balance between cost and quality? Design tends to be very much about focusing on designing that service in the context of where the user will mainly interact with it. But as we all know, the relationship with a new product, be it a car or a TV program, starts and ends well before you sit inside the car or view the program. Whether the customer is motivated to continue to use and pay for the service and recommend it to friends or colleagues is largely determined by this wider experience.

A typical checklist of requirements using a holistic approach to user interface design could be as shown in Table 3.2.

All the aforementioned items create the whole user experience. It is not just the words and the graphics on the screen, which are all too often the focus of usability. Most successful projects depend on the ability for marketing and software development departments to have a good working relationship. If usability only creates constraints for the designers' proposed user interface, it degenerates into the process known as interface policing and is a stumbling block to making a good design.

3.2.3 Designing for tasks

Throughout the design process, always keep in mind what the user is trying to achieve with the service. In effect, you have to get inside the head

Table 3.2
User Experience Checklist

What is the learning process?	How will the user find out how to use the service?
What will be the support process?	What help will be available? This could be either on-line or via telephone support using the call center (if one is available).
What is the documentation?	Maybe a user manual, quick reference card, or on-screen instructions.
What's in the box?	The contents of the standard package (e.g., phone, battery, charger, SIM card, and welcome pack).
What is the out-of-box experience?	This can be either a plug-and-play service that can be performed by the customers at their leisure, or a preconfigured package at the point of sale.

of the user. One way of doing this is by creating written scenarios or use cases of how a system will be used. While those scenarios don't have to be verbose, they should be written in a realistic manner. The character involved should have a name, profile, profession, etc. Then document how that user would access the proposed service in a typical day. If you create a number of profiles, they can be used as a guide in the design process. Referring back to them can provide commonsense ground rules and aid in testing at a later date (see Table 3.3).

There is obviously no end to how many of these scenarios one could invent. However, it will quickly become clear that the same design opportunities arise again and again. Here, we can see that all three scenarios start with checking the current account balance, and then the user makes decisions based on that. Our service should probably be designed so that this information is always displayed up front, with the other option—statements and transfers—available from there.

Table 3.3
Sample Set of Scenarios for a Banking Service

User	Eric Smith; Forty-five years old; Marketing Manager; Little computer literacy; Holds on-line checking account, savings account; Frequent traveler.
Scenario 1	Eric received a bank statement this morning, which did not include some recent items he was expecting, and he wants to check what has happened. He uses his WAP service to check the balance of his current account. He still is not sure everything is up-to-date, so he checks the recent transactions and can see that one check has not cleared. He decides he now has a clear picture and can wait another day or two.
Scenario 2	Eric is moving and needs to be absolutely sure he has funds to cover a substantial check. He is at home and has a choice of using his PC, or calling the bank's call center or his WAP phone. The WAP service was so convenient the last time he used it that he chooses it now to check the balance of his current account. It can barely cover what he requires, so he transfers some money from his deposit account.
Scenario 3	Eric is shopping with Mrs. Smith. In the boutique, he checks his balance while she is trying on some clothes. It is a bit higher than he thought, and he decides to buy himself some new clothes as well.

3.3 Design principles

The great benefits of WAP services—accessibility, compactness, and ubiquity—present challenges for the design of the user interaction with the service. The following principles should guide how WAP service's user interface design can address these challenges.

3.3.1 Economy

The user's terminal channel of communication is narrow—the terminal can only convey a little information at a time, and the user can only enter simple data. It is intrinsic to the type of user and service—so much to do, so little time—that a task must be completed with a minimum of interaction. This can be achieved by two principal means:

1. Adopting a holistic approach to involve the right parts of the organization and to concentrate on the users' tasks. Good services are developed for and with users: the needs of the user are paramount at all stages.

2. Using personalization to minimize input—see Personality (Section 3.3.3).

3.3.2 Modularity

Within a single service, and across a range of services, we can identify common elements such as selecting a date, or even authorizing a credit card payment. These elements should be standardized, so that the user can easily perform the same task in different services, does not have to learn a new procedure each time, and can use any service with confidence.

3.3.3 Personality

A personal profile of the user—contact information, account numbers, user IDs, etc.—should be maintained by the service and used to adapt content to the user. The service should record data entered and choices made so that recent entries can be offered as defaults. Personality should also be an attribute of the service as well as the user: the personality of the service is the brand. Branding may be applied on two levels:

1. The umbrella service brand. This may be a mobile operator (see Chapter 8), a content aggregator, or some other brand. The goal is

for users to identify with the umbrella brand as a comprehensive and reliable source of mobile services.

2. The content provider may also brand services, presumably to tie into their existing nonmobile brand.

3.3.4 Synthesis

A common constraint on a WAP service is to create a seamless environment with a Web service. It differs from shrinking a Web site from a desktop environment to a mobile device because there is a level of cooperation rather than competition. A complete service to the end user which is available on the Web and WAP may have elements that are only available on the Web site, while other aspects may be accessible on the WAP front end. We ought not to see WAP service design as shrinking a Web site to a mobile device and producing a hybrid site that supports Web and WAP. What we should strive for is the consistent use of terminology, for instance, referring to objects (portfolios, accounts, and so on) with the same names and always using the same phrases for executing or canceling tasks. In the case of Digital Mobility's Inhand service, we use the Web front end for registration and service provision and configuring how the service should work for you (e.g., which banks, travel agents, and stockbrokers you want to use).

In order to create a successful service, it is important to ensure that the two interfaces (Web and WAP) to the system are in harmony, working in a synchronized, complementary fashion. As much as possible, there should be a positive transfer of learning. The purpose of a good user interface is to be able to provide services where the time of executing and learning the task is kept to the absolute minimum for novice users. Yet at the same time we want to be able to cater for expert users, who do not require the prompting and procedures laid out for the first-time user. One of the ground rules to creating a good user interface is to establish style sheets for both Web and WAP sites, so that a common goal is achieved, rather the two sites working at cross-purposes.

3.4 Input techniques

For information services, user interface design often concentrates on the display of content at the cost of user input technique. For example, many Web search engines provide boolean logic facilities with complex rules about whether "and" means "I want to find pages with both these words" or "I want to find pages with a phrase that includes 'and.'" It might have been better to provide separate boxes for each word.

As a case study, let us assume we are designing a timetable lookup service. The user has to choose where they want to go to and from and on which day. The service will then find the available planes, trains, or other modes of transportation. Let us also assume that there is a set of short codes for departure and destination points such as airport designators. It is essential that it is easy for the user to choose these points with a minimum of interaction.

A starting point, without much thought, might be a text prompt, a WML <input> element for entering a code and an <anchor> offering a link to a further card for looking up names (e.g., New York) to find a code (JFK). Depending on the device, this might be displayed as shown in Figure 3.1.

Let us reflect on this in light of our design principles and examine some improvements we could make.

3.4.1   Avoid text entry

Users will often make multiple searches using similar criteria, when they are considering alternatives, or even when they make exactly the same search later because they have forgotten the results. (Note that this statement is an assumption for the sake of argument: it should really be deduced from analysis techniques such as use cases and scenarios.)

However, on a typical WAP device like a phone, even entering short three- or four-letter codes is a time-consuming process. It may take a click to enter text entry mode, then up to four clicks to enter each letter and a final click to leave text entry mode: 10 clicks in total, perhaps.

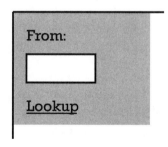

Figure 3.1   Departure selection—first design.

This is contrary to the economy principle, and we can look to the personality principle for a solution.

3.4.2 Defaults

We can extend our design to record the last choice made for the input field and enter that as a default next time the field appears. For example, see Figure 3.2.

Now, in the case where the user wants to use the same departure point, he or she has avoided 10 clicks. This is some progress over the first design.

3.4.3 Lists

Most travelers tend to travel regularly between a handful of points, so they will probably get to know the codes for those places, and will only occasionally need to use the Lookup facility to find others. However, 10 clicks to reenter each of two alternatives that the user may be considering are still too time-consuming.

We can use more personality to record a list of recently used choices. The exact number is obviously a tradeoff with screen space. However, let us assume that about five is appropriate, as shown in Figure 3.3.

Now, in many cases, the user can choose the departure point with a single click. (We assume that scrolling through a list of items is relatively easy.) The other facilities are still there to use, of course.

So, with the use of a bit of effort on the server side to remember recent choices, we can greatly improve the economy of the interface. However, this departure point element is no use on its own: it must fit into a navigation model.

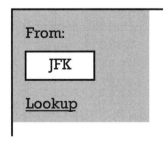

Figure 3.2 Departure selection—with default.

Figure 3.3   Departure selection—with list of most recently used.

3.5   Navigation models

Referring back to our immediate reactions, it's easy to think that the constraints of the WML specification and the small form factor do not allow the designer any freedom in choosing a navigation model. However, this is a lazy response, and we can now look at two different approaches and see how they affect a concrete example.

Suppose we are designing a timetable lookup service. The user has to choose where he or she wants to go to and from and on which day. The service will then find the available planes, trains, or other modes of transportation. Bearing in mind "three clicks and you're out," we want the user to be able to form these queries with a minimum of interaction.

3.5.1   Form-based navigation

By form-based, we refer to the normal dialog box behavior in Windows. For example, if we want to change the format of some text, we open a dialog that shows all the options available, select them in any order, and then press OK to apply them. Our timetable query might look like Figure 3.4.

We assume that the query is actually requested using a mechanism like a WML element with a "Find" label, for example. Note that we haven't heeded our previous advice about offering lists. We will address this later. Also, we have implicitly introduced a modular component: the

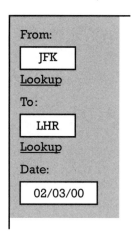

Figure 3.4 Timetable query—form-based design.

departure and destination points are chosen using the same set of WML elements.

Is this a good design? One good point is that the data items can be entered in any order, which may be important in some services. Also, however we enter all the details, we can review them all before requesting Find and change any with which we are not happy.

On the negative side, though, we have quite a lot of content on a single card. A typical phone would not be able to display content without scrolling, and the user would not be able to see all the data items at once. The worst feature is the number of clicks introduced to support navigation.

3.5.2 Question-and-answer navigation

This is a style of navigation similar to the Windows wizard. Basically, we break a task—in this case a timetable query—into a series of simple questions and answers. Our query now becomes a sequence of cards: see Figure 3.5.

Now, choosing an item from the most recent choices on the departure point card navigates immediately to the destination card. In this case, the user chose LGW, which is displayed in the fixed text as a confirmation. This is easily achieved using a WML variable (see Chapter 2 for details on WML variables). Finally, choosing one of the dates makes the query. Three clicks for the entire query!

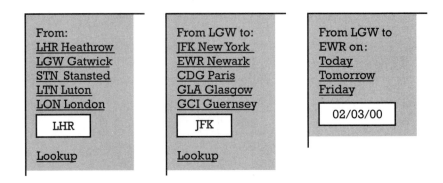

Figure 3.5   Timetable query—question-and-answer design.

3.5.3   Put the user in control

When we use a multicard navigation model like the question-and-answer design, we are keeping the user informed of the choices they have made. We ought to provide a back to let them return to the previous choice to correct it. Indeed, this is a golden rule: we should always let the user get back to where they came from.

Which navigation model is best? Well, in this case, the question-and-answer model appears to be more economical than the form-based one. But that conclusion is based on many hypothetical assumptions about the service and its users. In designing a real service, we should adopt the design processes we have already outlined. Only using techniques like scenarios and use cases can we analyze candidate designs to see which might work best.

The form-based model may work better for situations where a persistent set of data is being modified. For example, on a mobile handset, the settings—for ringing tone, volume, etc.—may be presented as a form. There is no need to navigate through settings we don't want to change, and there is no fixed end point to the interaction. In contrast, when sending a short message service (SMS), a navigation model more like the question-and-answer model may be used—first entering the message and then the recipient number, and finally the message is sent. There's a natural flow to the task from beginning to end. Of course, many other navigation models are possible: the two presented are just examples.

At this time, it is worth making a few points about security. Suppose, once we find a flight, train, or other mode of transportation, the service allows us to book a ticket. Presumably, this involves a payment by credit card or some other financial transaction (see Chapter 10 on mobile financial services for more information). It's in everyone's interests that this transaction is only carried out by the appropriate person, so we'd probably need the users to authenticate themselves with personal identification numbers (PIN) or some other security mechanism. This procedure inevitably introduces further input and navigation: time and inconvenience for the users. However, the need for this only arises if the users find an appropriate journey. If they're just browsing to weigh their options, we don't really want to have to bother them with authentication. We could defer the login process until the first time any security-critical transaction is definitely going to be required. This can be carried forward for any further security-critical transactions in the same session, so that the user only needs to go through the authentication process once in a session. Security on a transport level (as opposed to the application level described here) is described in Chapter 7.

3.6 Testing the user interface

There are a range of ways of testing final or prototype designs. One way of achieving this can be through interactive user workshops, where potential users are allowed to move around the Post-it notes of the prototype system to suit their needs.

Another powerful way of testing is to get real users to trial the prototype system, testing the tasks that have to be done in the environment that the users would normally use, and try to emulate what will happen in reality. At the end of the day, you have to make sure that your design and development and rollout process involve the end users as much as possible; otherwise, you risk not addressing their needs. The important thing to be aware of in terms of analysis and design is not to rely on the thinking and requirements of the designer, because the designer might not be representative of the target audience. People sometimes design and develop the user interface with themselves in mind rather than the end user and thus often may create flawed systems.

The further you get down the design cycle, the more you have to fix and pin down decisions for your product. After thorough testing using both technical and user groups, you have to close the door on any further development and create version 1.0 of the user interface. However, despite the release of the software, a process of refinement should be taking place. Iterations in the design process will be slower and more controlled, but should still carry on. You should encourage user feedback by request or monitoring use, the results of which can be fed into later software releases.

Testing a user interface can become a rearview mirror approach. It is not an ideal way of testing, but it does help to identify mistakes that you should have tried to analyze and design out at the pen and paper stage. However, the later you leave changes in the design cycle, the more costly and time-consuming it becomes.

3.6.1 Different devices

With the continual stream of new WAP-based products being introduced to market, designers must always adopt a flexible approach to designing the user interface. You have to do rigorous testing with all the popular makes and models of mobile devices, meaning a close relationship with the manufacturers is essential.

Multifunctional devices such as a palmtop computer are a compromise compared to a single-function device such as a phone. It does the same job, but may not be as light or as good to hold as other designs.

Take, for instance, a kettle versus a stove. Both are capable of boiling water. But in order to boil water in the kettle, you simply put water in it and flick the on switch and the device will switch itself off once the task is completed. With a stove you have to find a pan, turn on the heat, and then monitor the water as it comes to a boil.

For some people the compromise is acceptable; for others it is not. Different people will want different things. With technologies like Bluetooth, it will liberate designers and allow them to create modular devices, which can be small and separate from one another, but able to work together harmoniously (e.g., separate display and input device).

There is always a place for each type of device. It is wrong to say that one device is better than another, because no single device will suit everybody. People will want different things and emphasis on their requirements. Therefore, manufacturers will bring out devices that hit different segments in terms of need, such as an emphasis on voice, data, or flexibility. Also, there may come a day when people will want to use more than one device. Hopefully, there will be a good range of devices that suit different needs, and there won't be an obsession with trying to cram in as much as possible.

A modicum of common sense should be applied, though, because the prolonged use of a product or service should eventually become intuitive. The sooner you are able to master the modes of interaction with that device, the sooner you can develop an interaction style sheet to ensure your user interface works the same way on different devices with varying controls and screen sizes.

For example, interactive arcade games with multiple players promise to be one of the killer applications over the WAP interface. If you think about how somebody plays a game, it is crucially important that it becomes completely intuitive what all the different buttons do and the modes of interaction with the game, so the focus isn't looking at the keyboard, figuring out the controls, but knowing that you need to zap that character on the screen instead.

3.7 Future developments

Some future developments may have a significant effect on the user interface design of WAP services.

3.7.1 First-class WAP services

From the point of view of the user interface designer, perhaps the most important future development is the promotion of WAP services to first-class status on devices. That is to say, the reduction or elimination of the distinction between WAP-based services and the native services on the device—telephony and contact management on phones and applications on palmtops, for example.

Some developments that may enable this promotion to first-class status are shown in Table 3.4.

3.7.2 Adaptive user interfaces

There is a clear distinction between adaptable and adaptive user interfaces. Adaptable user interfaces are ones that you can configure and are common on phone devices where you can allocate speed dials, customize the menu system, and access shortcuts. This is good, because all people are individuals and have different requirements and prefer to do things in a different way.

The adaptive user interface is something that is significantly more complex and difficult to do. This is where the system monitors the usage pattern by the user and tries to adapt to what he or she is doing. The obvious danger is that human beings are extremely complex devices in their own right, and it is very difficult to predict what we, as humans, really want. There then becomes the danger that an adaptive interface becomes

Table 3.4
Developments to Enable First-Class WAP Services

Integration with device facilities	If a WAP service is able to use native facilities—such as using a built-in calendar to select a date for a travel service—the user will experience a much more seamless interface.
Predictive text entry in fields	Although many devices have predictive text entry for their built-in functions—SMS messaging in particular—a WAP service cannot currently make use of this.
Always-on connection	The inherent delay—of the order of 10 to 30 seconds—to establish a circuit-switched connection to a service is a significant discontinuity and a potential disincentive to the user. Always-on technology such as the general packet radio service (GPRS) will overcome this.

a nuisance because the system is trying to force you into doing things in a way that it thinks is helpful, but in fact is irritating.

Despite these difficulties, adaptive interfaces are something worth looking at. When you are in the mobile domain, you have these restrictions of screen size, weight, and so on. So anything you can do to assist the user in speeding up input and output has got to be an advantage. It is always something a designer should be thinking about. Are there things that we can monitor by frequency of use or by input values to which you can then default? It is something that has to be used carefully, as it can become an irritant, but when it works well, it is brilliant. Do you feel that the system works with you?

A way to ensure that any part of the system that has adaptive capabilities prevents becoming an irritant is the ability to switch these settings off when required.

Some people say that adaptive interfaces are an impossible task. Human beings are too complicated. But the potential benefit is great and is worth at least considering, even if you end up rejecting it within a design.

3.8 Conclusions

This chapter has touched on some popular approaches to creating a good user interface. While it does not aim to offer a definitive explanation to the ins and outs of user interface design, it will provide valuable pointers to crafting a sturdy and reliable platform for delivering an effective WAP service.

An essential checklist of questions that Digital Mobility might ask when building a WAP user interface for a service could be as follows:

1. Who are the users?

2. What are the essential tasks?

3. Can new users understand what to do before they start?

4. Is security at the right level—secure, but not intrusive?

5. Are there defaults wherever possible?

6. Is data reentry avoided at all costs?

7. Can the user always back out of a task?

8. Does the WAP channel complement the user's existing channels?

9. Does it use consistent terminology?

10. Three clicks and you're out!

Contents

Wireless Telephony Application: Telephony in WAP

Magnus Larsson

4.1   Introduction

WAP introduces the mobile-device user to a world of new services. You can imagine what will be available to WAP-enabled devices in the near future by just looking at services accessible on the Internet today. If there is anything you need to know about currency exchange rates, up-to-date stock quotes, train timetables, articles and news, etc., it is available once you have the address to the information provider and a browser through which the information can be retrieved and viewed. Of course, services that are available through WAP have to be adapted for the WAP environment, but this probably represents a cost that is much lower than the revenue that can be expected by more frequent use of these services and of the underlying network bearers.

But WAP is not only about collecting information from content providers and presenting it on a mobile device's screen. WAP also defines a framework that enables content authors to use telephony features from within mobile telephony devices. This particular toolbox is called the wireless telephony application (WTA) framework. WTA allows for applications written in WML and WMLScript to interact with a mobile device's telephony-related functions. Consider the following example.

Let's say you need some help with the plumbing in your house. In your WAP-enabled phone you have previously stored a bookmark pointing at a "yellow pages" service somewhere on the Internet. This service lists telephone numbers to businesses in your neighborhood. You can use the browser to enter a search criterion, which in this particular case probably would be "plumber," in order to get the address and telephone number to a professional nearby. A list then appears on the device's screen. In this list the names and numbers are underlined to show that they can be selected in some way. Now, the provider of this list has used the WTA framework facilities to associate these names and numbers with functions that interact with the call setup features in the device. Thus, when you have scrolled the listed numbers and finally selected one, the phone will set up a call to the plumber of your choice and you can discuss which pipes have to be replaced.

This is a typical example of how a third-party content provider can offer a WAP-device user services that only need the basic telephony feature of setting up a mobile-originated call. The WTA framework defines this very restricted but useful feature as being "public." An authorized provider of advanced telephony services will be able to use the more extensive set of "network" features. These include generic interfaces to call management, phonebook, and network text functions accessible from within the mobile device. This part of WTA also allows for events, caused by changes of state in the mobile network, to be associated with content stored in the WAP client. The reason for having such content stored locally in the device is to avoid the delays that would be caused by retrieving the content from a server each time the associated event occurs. The WTA framework specifies a mechanism for downloading that kind of content from a content server.

Essentially, the WTA framework features transform a microbrowser environment, with means to exchange WML-based content with a server, into a platform for executing services that connect the information space supplied by the WAP domain with mobile network services provided by the mobile telephony service provider. With the presentation facilities offered by WML, the dynamic behavior possible by using WMLScript, and the features supplied by the WTA framework, a telephony service provider can write advanced, interactive services that are presented to the WAP-device user in an appealing manner.

This chapter aims to go a little deeper into the WTA framework and to give detailed descriptions of each of its components. In the last section you will find an example that shows what a service that uses some of the advanced network features could look like.

4.2 WTA architecture overview

WTA [1] is an extension to the wireless application environment (WAE) framework [2, 3] (see Chapter 2 for a description of the elements of the WAE). It assumes the same architecture model as WAE, comprising an origin server, a WAP gateway, and a client. Figure 4.1 displays the WAE architecture.

An origin server stores content statically, or generates it upon a request from the client. The client requests content or services from the

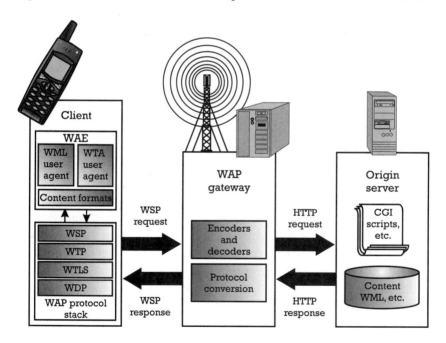

Figure 4.1 WAE architecture overview.

origin server by using URIs that point to the location of the content. The URI scheme used is the same as in the Internet world. Translation of WAP protocols (wireless session protocol—WSP, wireless transport protocol—WTP, wireless transport layer security—WTLS, and wireless datagram protocol—WDP) to Web protocols (HTTP, SSL/TLS, and TCP/IP) is performed by the WAP gateway, as well as encoding and decoding of WML [4] and WMLScript [5] content when transferred from and to the origin server.

The initial work on WTA within the WAP Forum has been focused on the client side. For the server side it is assumed that there will be origin servers with means to interact with entities within the mobile network. This kind of server is referred to as a WTA server.

4.3   The WTA framework components

The WAE framework supports an application architecture model where different kinds of WAE user agents have capabilities to process well-defined services and content formats (see Figure 4.1). A WML user agent, for instance, handles WML and WMLScript elements. The WTA framework extends this model by adding a WTA user agent and three WTA-specific features on the client side: a WTA interface, a repository, and an event-handling mechanism. Figure 4.2 shows the client with the WTA framework components.

4.3.1   The WTA user agent

A WTA user agent is one type of WAE user agent that the WAP-specification suite references. Another one is the WML user agent. The WTA user agent is a WML user agent with extended functionality. Key features of a WML user agent are the ability to render and execute WML and WMLScript content retrieved from an origin server via the WAP gateway. A WML user agent also supports an event-handling mechanism for binding tasks to events originating from a user's interaction. A user can interact by navigating between WML decks, or between cards within an active deck, or by selecting an option in a list, thereby causing so-called intrinsic events. Tasks bound to these events could be automatic navigation to other cards, or invocation of WMLScripts, etc. See Chapter 2 for more details on WML features.

In addition to these capabilities, a WTA user agent has access to all WTA framework features. Hence, the WTA user agent can also use the

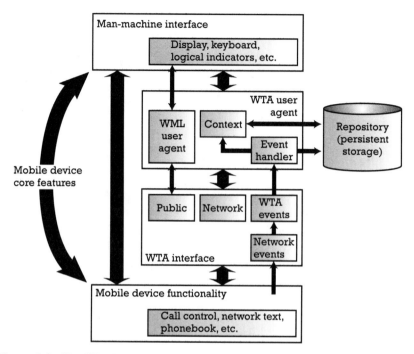

Figure 4.2 The WTA framework components.

mobile device's telephony functions through a WTA interface (WTAI), store WTA service content persistently in a repository, and start WTA services by handling events originating in the mobile network and associating them with content in the repository. A WML user agent cannot access features in the WTA framework, with one exception: content executing in the WML user agent is allowed to invoke a limited set of WTAI functions (described in Section 4.7.2).

4.3.1.1 WTA user agent context

When discussing WML or WTA user agents the term "context" is often used. A context is an abstraction of an execution space that holds all current parameters, navigation history, and state of the user agent. A service that is executing in the user agent is a part of the context, and defines the extent of the context. A context can be cleared when the user agent needs to be set to a well-defined state, which would be appropriate when execution of content that is part of a new service is about to begin. When a context is cleared, all navigational history state is erased and content associated with the current service is removed from the user agent's active memory. A new and clean context is always initiated when the WTA user agent is started. When the WTA user agent or the current context terminates, an executing WTA service will end. However, WTA offers a service the means to define whether phone calls established within that service's context should be terminated at the same time, or proceed until any of the parties hang up.

4.3.1.2   Use of WSP session services and WTLS

To interact with a WAP gateway, a WTA user agent can use both connection-mode and connectionless-session services offered by the WSP [6]. WSP session services requested and used by the WTA user agent are always established on specific, secure WDP [7] ports on the WAP gateway. These ports identify the WTLS [8] as the layer to be used above the WDP layer between the client and the WAP gateway. The first releases of WTA assumed that WTLS class 2 is used, which implies that content transferred between the client and the WAP gateway is encrypted and that the client can authenticate the WAP gateway via the gateway's certificate, stored in the client. Authentication of the WAP gateway is essential for enforcing security in WTA. See Chapter 7 for a detailed description of the WTLS.

4.3.2   The WTA interface

Creating telephony services requires access to telephony features. The WTA interface is a generic and high-level interface to a mobile device's internal telephony-related functions. Among these are functions for setting up phone calls and reading and writing entries in a phonebook.

Content that executes in a WAE user agent will have access either to a restricted part of WTAI, which is the public WTAI [9], or to the whole set of functions, represented by network-common and network-specific WTAI [9–12] (see Section 4.7 for a description of these function categories). Which level of access is given depends on who is providing the content, a trusted or not trusted WTA content provider, and whether the content is executed by a WTA user agent or not.

4.3.2.1   Access to public WTAI

A WML user agent can retrieve any kind of content from any location when used for browsing the Internet. Giving a WML user agent access to the whole range of WTAI functionality could allow malicious use of the mobile device's telephony functions. Still, a reliable third-party content provider should be able to offer the WML user agent user to download and execute content that uses basic telephony functionality. For that reason a WML user agent is allowed to use the public WTAI. At present, public WTAI includes two functions: one for setting up a mobile-originated call, and one for sending DTMF tones over the established call connection. At the time the functions are about to be invoked, the user has to be notified in order to grant the service permission to use them. This is a security measure applicable to all WTAI functions in order to have the user decide what manipulation of the device is allowed.

Public WTAI does not include mapping of network originating events to so-called WTA events (see Section 4.7.6). A service executing in a WML user agent can therefore not be triggered by network events.

4.3.2.2 Access to network-common and network-specific WTAI

Network-common and network-specific WTAI define the second level of access to in-device functionality. These categories offer the complete assembly of WTAI features, including the interface to WTA events. Services delivered by a WTA service provider (defined in Section 4.5) have full access to network-common and network-specific WTAI when executed by the WTA user agent. However, as is the case with public WTAI, a user must have the possibility to decide whether a service should be given permission to invoke functions in network-common and network-specific WTAI. This kind of user permission can be called upon each time a service invokes a function, or the service can be given a blanket permission to use that function whenever the service executes.

4.3.3 The repository

A WTA user agent uses the WAP protocols for requesting WTA service content from a WTA server. When designing the WTA framework, one requirement was the WTA user agent's ability to support timely execution of specific content in response to an event originating in the mobile network. As an example, an event in the form of an incoming call indication has to be acted upon immediately to not give the calling party a reason to hang up. The conclusion was that this content had to be stored on the client side, since requesting it from a server each time the event occurred would introduce unacceptable delays and unnecessary use of the narrow network bearer.

The repository, described further in Section 4.8, is a persistent storage container that aims to fulfill the real-time requirements on WTA services. The repository model includes the definition of the WAP channel content format that is used to define a WTA service. Channels and their content are delivered from a WTA server upon request from a WTA user agent, and stored in the repository. Storing content persistently in the client also gives the WTA user agent the advantage of not having to rely on a session established to the origin server at the time of the event.

4.3.4 An event-handling mechanism

WTA services stored in the repository can be initiated by a WTA user agent user, provided the WTA user agent is somehow presented with a menu list of the available services. But WTA also defines a mechanism for automatically starting WTA services in response to events generated in the mobile network. A global event binding (defined in Section 4.9) associates a WTA event, which is a network event transformed to a format suitable for the WTA user agent, with a service stored in the repository. An incoming call indication could, for instance, start a service that displays a list of possible actions for the user: answer the call, reject the call, or forward the call to a specific number.

The event-handling mechanism also supports events that occur within an executing service. After the initiation it is likely that a service expects new events to happen. An executing service captures events by defining temporary event bindings (also defined in Section 4.9). A service that is intended to capture an event has to define a temporary binding to that specific event, as well as the action to take upon reception of it.

4.4   The WTA server

A WTA server is an origin server holding content that can be requested by a WTA user agent. The content on the WTA server is referenced using URIs. In that sense it is not different from any other origin server. What makes a WTA server unlike other origin servers is that it also interacts with other mobile network components. A voice mail server could be such a component (see Figure 4.3). Since URIs can be used to reference applications as well as content on an origin server, a WTA server has the means to interact with entities in the mobile network upon a user's request for a URI. A WTA server that is somehow connected to a voice mail system, for instance, can get information about the user's status regarding received voice mails, and inform the user via WAP and the WTA user agent.

4.5 WTA services and WTA service providers

A WTA service consists of executable content that uses the features provided by the WTA and WAE frameworks. Content building a WTA service is typically stored in the repository and triggered by events in the mobile network, using the event-handling mechanism defined in WTA and accessing the mobile device's functionality through WTAI.

A WTA service is delivered by a WTA service provider, who could be the mobile telephony service provider (the operator) to which the user subscribes, or a content or service provider that is authorized by the mobile telephony service provider to deliver WTA services. A WTA service provider offers enhanced telephony services to a WTA user agent by providing content and services accessible on a WTA server.

4.6 WTA security model and access control

When transferred from a WTA server to a client, WTA service content is separated from other content by the use of different WDP port numbers on the WAP gateway. A WTA user agent always uses specific WDP ports on the WAP gateway when establishing a WSP session, and such a session is the only one allowed to transfer WTA content to a WTA user agent. Content that is not related to WTA services is to be transferred through the WAP gateway using other predefined ports. This mechanism is pictured in Figure 4.3.

The security mechanism presently available in WAP provides transport layer security. This security is implemented using WTLS between two WTLS connection endpoints of which a client is one and a WAP gateway, or an origin server with built-in gateway functionality, is the other. WTLS allows for the WTA user agent to authenticate a WAP gateway and have WTA service content encrypted when transferred between the WAP gateway and the WTA user agent. A WTA user agent uses this authentication to identify specified gateways that are supervised by the mobile telephony service provider and trusted for delivery of WTA services. At the time of writing this chapter (early 2000), there is no standardized mechanism defined in WAP for delivering the identities of these trusted gateways to a client. There is, however, work going on to specify how provisioning of such information should be done.

To extend the chain of trust beyond the WAP gateway and to the WTA server that delivers the actual WTA services, the WAP gateway, or

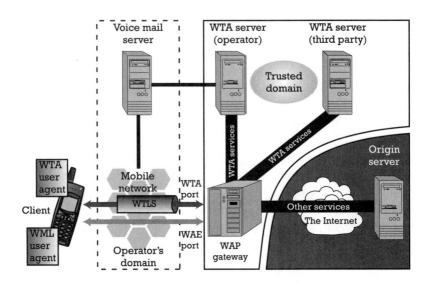

Figure 4.3 Security model and access control.

its supervising telephony service provider, must ensure that there is a trust relationship between the WAP gateway and the WTA server. Only a WTA server managed by a WTA service provider is approved to access the trusted gateway. How this trust is achieved or what technique should be used to enforce security between these entities is up to the mobile telephony service provider. It might be appropriate to use SSL/TLS, the protocols from which WTLS is derived.

This solution does not provide end-to-end security since it resides on the transport layer level, and the WAP gateway has to translate between protocols when transferring content. Content is thereby revealed to the possessor of the gateway. This is probably not a problem when the operator guards the WAP gateway. But there might be other solutions where security has to be maintained even if the WAP gateway is not trusted. The WAP Forum is currently driving several efforts to define end-to-end security solutions. When completed, these will be a part of the WAP overall framework and available to application frameworks such as WTA.

4.7 WTAI—interfacing WAP with the mobile network

4.7.1 The WTA interface design
The WTA framework is targeting mobile devices that have built-in functionality for managing phone calls. Some of these devices also have

capabilities for sending and receiving text messages, maintaining call logs, managing phonebooks, etc. The set of features available in a device is, of course, due to a choice made by the device manufacturer, but the fact that each device is designed to conform to one or more mobile network standards also affects the actual device capabilities. Different network types have different characteristics, and therefore the services offered to mobile devices and their users may vary.

A WTA user agent executes telephony applications. To be able to create telephony applications based on WML and WMLScript, there is a need to have access to the telephony-related functions in the mobile device. The answer to this need is the WTA interface, WTAI.

WTAI is a collection of functions forming an interface to features in the mobile device. Some of these features are truly in-device, such as a phonebook or a call log. Some of them cause the device to interact with the mobile network services. One example of the latter is the feature of setting up a mobile-originated call. Events that derive from changes of state in the mobile network are mapped to WTA events, also defined as a part of WTAI.

The WTAI functions are grouped into libraries, all functions in a library being related somehow. For that reason there is, for instance, a phonebook library that offers an interface to the in-device phonebook. There is also a voice-call control library encompassing functions for setting up a mobile-originated call and terminating it. Figure 4.4 presents a view of the WTA interface.

The different libraries are divided into three categories: network-common WTAI, network-specific WTAI, and public WTAI. These categories reflect the availability of the functions in the sense that some functions are generic and applicable in all mobile devices and networks, and some are only relevant in specific networks. Hence, functions defined in network-common WTAI are those that apply to all mobile networks. These encompass the basic telephony functions. Functions specified as being part of network-specific WTAI are only available in a device conforming to a specific network type. At present, there are three network-specific libraries: the GSM, the PDC, and the IS-136 specific libraries. Functions defined in the network-common and network-specific libraries are only accessible from applications provided by a WTA service provider and executing in the WTA user agent.

Public WTAI, which is the third category, is designed for non-WTA applications, meaning applications that are not provided by a WTA service provider or not executing in the WTA user agent. In principle, any

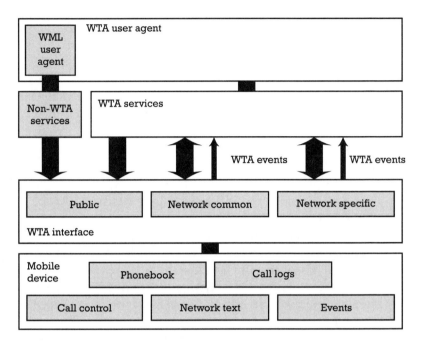

Figure 4.4 WTA interface to mobile-telephony device functionality.

third-party content provider can build services that use public WTAI. Therefore, public WTAI functions are a small set of basic and generic telephony functions with limited possibilities to manipulate the mobile device.

4.7.2 Public WTAI

All public WTAI functions are gathered in one library. These functions are made available to any WAE user agent, of which a WML user agent is one example. The reason for having public WTAI is that it can be used by applications that are not allowed, or do not need to have access to more than basic telephony functions. The public library defines two functions, namely *make call* and *send DTMF tones*.

The make call function allows the application to set up a call using a valid telephone number. Since there is no corresponding function in public WTAI for terminating a call that has been set up by this function, the mobile device's core functionality has to be used. Send DTMF tones is meant for use in conjunction with the make call function. A sequence of standard DTMF characters is passed to the function, resulting in the corresponding DTMF tones to be sent through the previously setup call connection.

With these two functions it is possible to write an application that, for instance, retrieves a phone number during a browsing session in a WML user agent, then lets the user set up a call using that number and maybe send DTMF tones for selecting a service once the call is connected.

4.7.3 Network-common WTAI

The network-common WTAI is divided into five libraries that are a lot more extensive than public WTAI. All of the functions that are part of network-common WTAI are available to a WTA user agent.

4.7.3.1 Voice-call control library

Functions included in this library are *setup call, accept call, release call,* and *send DTMF tones.* These are functions that should cover all basic features needed by an application for managing phone calls. The caller of the functions setup and accept call can decide whether a phone call shall be dropped or kept if the context in which it was set up, or accepted, is terminated before the call is ended.

4.7.3.2 Network text library

This library provides the application with *send text, read text,* and *remove text* functions. The purpose is to present a WTA user agent with an interface to the mobile telephony device's messaging functionality. In GSM networks these functions map to short message services (SMS).

4.7.3.3 Phonebook library

Access to the device's phonebook is accomplished by offering an interface with the functions *write phonebook entry, read phonebook entry,* and *remove phonebook entry.* When reading the phonebook, it is possible to search the entries for a match with an entry identity, a name, or a number.

4.7.3.4 Call logs library

Most devices store a history of phone calls in call logs. Such logs are made accessible through the functions *last dialed numbers, missed calls,* and *received calls.*

4.7.3.5 Miscellaneous library

Functions that are not easily identified with a specific library are collected in the miscellaneous library. These functions are not necessarily interfacing the mobile device, but are used for context management within the user agent. The functions defined are *indication, terminate WTA user agent,* and *protect WTA user agent context.*

The indication function is used to turn a logical indicator of the mobile device on and off. The caller of the function cannot decide how an indication should appear in the device MMI, only what type of indication is requested. For instance, the device can be asked to present a notification of the type of incoming speech call. It is then up to the device implementation to choose the appropriate indicator.

The terminate WTA user agent function provides the content author with the possibility of terminating the context currently executing in the WTA user agent. In effect, this means that all navigational history state is cleared and all content associated with the current WTA context, including variables, is removed from the WTA user agent's active memory. Calls that are active when this function is invoked are dropped or kept as requested when they were set up.

It might be of importance that an executing service is terminated only by the user or by the service itself, using the terminate WTA user agent function. However, due to the event-handling mechanism in WTA, an executing service could actually be terminated as a result of a received WTA event (see Section 4.9.2). In such a case, the protect WTA user agent context function could be used. If the service protects itself, it will not be terminated as a result of a received WTA event.

4.7.4 Network-specific WTAI

Even if the network-common functions represent a large part of the functionality present in mobile networks and devices in general, there are still features that are not covered and must be defined for each specific network. At present, WTAI has network-specific extensions for GSM, IS-136, and PDC networks. Each network type has its own library of functions.

4.7.4.1 GSM specific library

GSM networks have an extensive set of features for call handling. Some of these functions are made visible to a WTA application through the functions *call reject, call hold, call transfer, join multiparty, retrieve from multiparty, provide location information,* and *send USSD.*

4.7.4.2 PDC specific library

PDC has features similar to GSM and offers the functions *call reject, call hold, call transfer, join multiparty*, and *retrieve from multiparty*.

4.7.4.3 IS-136 specific library

IS-136 networks use flash and alert codes, and these are made accessible with the functions *send flash code* and *send alert code*.

4.7.5 Calling WTAI functions

Each WTAI library has a unique name. The same applies for all the functions in a library. So, the function name preceded by its library name forms a unique reference to a WTAI function.

Telephony applications can be written using both WML and WMLScript. For that reason, WTAI functions are made accessible from both. The WMLScript interface to a function consists of the library and function names, dot separated, and followed by the parameters to be passed to the function. Setting up a call to a party with the number 23456789, using the public function *setup call*, would look like this in WMLScript:

```
WTApublic.makeCall("23456789");
```

Accessing WTAI functions from WML requires a specific URI scheme. WTA specifies such a scheme. The library and function names in a WTAI URI use a short form of the names used for the WMLScript interface. The same function call as the one previously described would look like this when using the WTAI URI:

```
wtai://wp/mc;23456789;
```

WTA defines a set of error codes to be returned by network-common and network-specific functions in case the functions would fail for some reason. What could happen is, for instance, that the end-of-a-call log has been reached due to consecutive read operations using the WTAI function received calls. In that case, the WTAI function returns an appropriate error code and the content author would be able to define proper actions.

4.7.6 WTA events

The network, to which the device is connected, delivers notifications to the device as a result of network signaling. For instance, an incoming call will generate an event to the mobile device and thereby allow for the device and its user to act upon that event. This way a call can be answered.

It is in the nature of WTA services to be aware of, and able to act upon, events originating in the mobile network. Therefore, the WTA framework specifies a set of so-called WTA events that map to these mobile networks' native events. An incoming call event generated by the network will be transformed to a WTA event that in turn triggers a WTA service.

The WTA events considered, being common for all network types, are part of network-common WTAI. At present these events are *incoming call indication, call cleared,* and *call connected.* Events that are only generated in specific network types are defined in network-specific WTAI, and so the *USSD message received* event is specified for GSM, and *incoming alert info* and *incoming flash info* events are defined within the IS-136 extension to WTAI.

4.8   Repository

4.8.1   A persistent storage for fast service access

The repository serves as a container for content related to WTA services. The reason why content should be stored locally in the device is to minimize transmission of data over the mobile-network bearers. The networks used are relatively narrow banded and the shuffling of content could be time-consuming, especially when compared to what is acceptable in interactions between a user and a telephony service. It is understood that all users want fast access to the service requested and that transitions between phases in a service must progress without unnecessary delay. The repository is aiming to fulfill these real-time requirements on WTA services.

4.8.2   Channels and resources

WTA specifies a content format, the channel, for defining WTA services that are stored in the repository. A channel document specifies an identity by which the service is referenced and a set of resources that implement the service. Each channel can reference a number of resources, and different channels can share the same resources. All the listed resources in a channel are to be downloaded from the WTA server and stored in the repository before that service can be referenced from within the

WTA user agent. Figure 4.5 describes the relation between channels and resources.

The channel is an XML [13] document that is delivered from the WTA server to the WTA user agent. It comprises four elements: the *channel*, the *title*, the *abstract*, and the *resource* elements. The channel element is the "head" element whose content is built up by the other three.

4.8.2.1 The channel element

The channel element is the "entry" to the service defined by the channel document. It has five attributes describing it. The `EventId` attribute is used when the service defined by that channel and its content is to be executed. This happens when a WTA event is matched with the channel, (i.e., there is a global binding from an event to the specific channel (this process is described later in Section 4.9)). It could also be that a user is able to reference a channel through an implementation-dependent presentation, in which the channel's title (see Section 4.8.2.2) element could be of help.

A channel that is meant to have a binding to a WTA event uses a specific naming scheme: the `EventId` attribute value must have the format `wtaev-xx`, where `xx` is the abbreviated form of the actual WTA event.

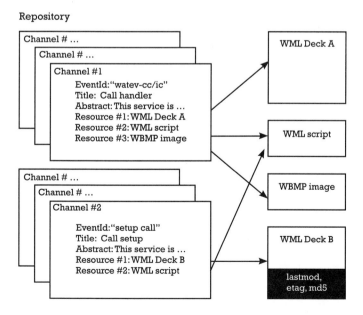

Figure 4.5 Channels and resources.

For the Incoming Call Indication event to bind to a channel, that channel would have the `EventId` attribute value set to `wtaev-cc/ic`.

The `maxspace` attribute indicates the total size of the channel and its resources. This attribute is used when a channel is about to be downloaded to the repository, in order to prevent unnecessary download of channels that will not fit into the repository anyway. The `base` attribute is a URI, which points to the location from which the channel content, the resources, is to be fetched when downloading it to the repository.

There are two attributes used for communicating successful or failed channel download. These are one `success` URI and one `failure` URI. Upon successful download, the WTA user agent requests the `success` URI from the WTA server and thereby notifies the server that the channel has been updated in the repository. If the download fails for some reason, the `failure` URI is requested.

4.8.2.2   Title and abstract elements

There are two descriptive elements in the channel document. The `title` element is used to carry a human-readable title of the channel. It should be restricted in length in order to be displayed on the mobile device. The `abstract` element, which is optional, carries a human-readable description of the channel. The description should provide the user with information about the purpose of the channel.

4.8.2.3   The resource element

The channel element lists one or several resources that define the content that constitutes the service. The first listed resource element is a reference to the content that will be invoked and executed when a channel is referenced (through the `EventId` attribute). This first resource is the "main" resource and it invokes the other defined resources as appropriate.

The channel element attribute `href` is a URI pointing at the location of the resource and is referenced both during download and execution. That is, when downloading a resource, the URI points to the location from where it is fetched for storing in the repository. When a service, of which this resource is a part, eventually references the resource to execute it, the service uses the same `href` attribute.

The resource element further specifies three attributes, `lastmod`, `etag`, and `md5`, to be used for keeping track of its freshness. If a resource is updated on the server side, its attribute values will also be changed. The attribute values of the resource in the repository will then differ from the ones on the server side and this indicates that the resource in the repository should be updated.

4.8.3 Channel loading and unloading

The channel document and its referenced resources are delivered as a part of the response to a standard URI request initiated by the WTA user agent and sent to the WTA server. But download of a channel could also be server initiated if the server uses the WAP service indication [14] feature. In that case, the WTA user agent is presented with a so-called service indication that carries the URI to the channel to be downloaded from the WTA server.

The installation of a channel is executed in five steps.

1. The URI for the channel content location is presented to the WTA user agent.

2. The channel URI is requested by the WTA user agent, thereby causing download of the channel document to the WTA user agent. After that the WTA user agent knows what resources are part of the channel as well as their freshness.

3. The resources listed in the channel document are downloaded from the WTA server. Only the resources that are newer than the ones already stored in the repository are downloaded. If the channel document does not reference any newer versions of the resources, the downloading will terminate here.

4. Once all new resources are successfully downloaded, the WTA user agent requests the **success** URI from the WTA server in order to notify the server that the channel is about to be activated.

5. The server responds to the success URI. This causes the WTA user agent to store the new channel and the new resources in the repository, updating the resources' **lastmod**, **etag**, and **md5** attributes. Finally, the channel is activated, meaning that it is made visible to and accessible from within the WTA user agent.

If an error occurs during installation, then any previously stored channel is preserved, and the new channel is discarded. In this case, the WTA user agent requests the **failure** URI that notifies the WTA server that the installation failed. The server responds to the **failure** URI, but if the response does not reach the mobile device, the WTA user agent should present some kind of error message to the user. If a failure occurs and there is no `failure` URI present in the channel, such a message must always be presented.

4.9   Event handling

4.9.1   Event bindings

The handling of WTA events is one of the cornerstones in WTA. But how can a WTA event be captured? The mechanism used is called event binding, which is an association between an event and the content that should be processed upon reception of that event.

A network-generated event can occur anytime. The WTA user agent could already be in a state of executing a service that expects the event to happen, or the executing service might have no relation whatsoever to the event. There is also the possibility that there is no executing service in the WTA user agent. In all these cases it should be possible to capture the event. For that reason there are two kinds of event bindings: global bindings and temporary bindings.

A global binding to an event exists when there is a channel corresponding to that event stored in the repository. The event causes execution of the resources defined by the channel. Temporary bindings are those defined in an already executing service. The binding is temporary in the sense that it only exists when this service executes. There might, of course, be other services, but not executing at the time of the event, defining a binding to the same event. However, the executing service is the one that institutes the binding.

A temporary event binding overrides all other bindings to that same event. So, if an executing service and a channel in the repository both have bindings to a detected WTA event, it is the temporary binding that will be processed. The action to take when the event occurs is defined in the executing service.

In case the WTA user agent fails with starting a service that is bound to a WTA event, the current context will be terminated and the event is handed over to the mobile device's own functionality for handling.

4.9.2   Event-handling procedure

There is a set of rules defining in which order events should be matched with global or temporary bindings. This matching depends on whether a service is executing or not, and if a context has been protected (using the network-common WTAI function protect WTA user agent context defined in the miscellaneous library).

1. *No service is executing.* The most straightforward case is when there is no service executing. By default no temporary binding can be found (step 1 in Figure 4.6). The WTA user agent will then try to find a global binding to the detected event. If such a binding exists, the new service as defined by the referenced channel will be initiated (step 2 in Figure 4.6). If this fails, the event should be handed over to the mobile device's own functionality for handling that event (step 3 in Figure 4.6).

2. *Service is executing.* When a service is already executing, there are actually three possibilities:

   ‣ The executing service defines a temporary binding to the event and the defined action is taken (step 1 in Figure 4.7).

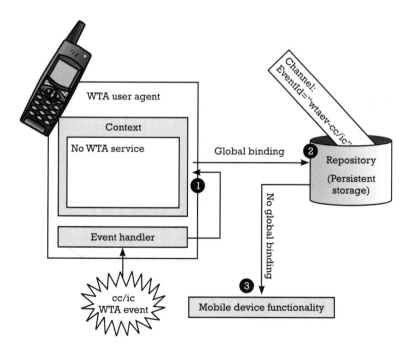

Figure 4.6 Event handling—no service is executing.

- The executing service has no binding to the event and the context is protected. In this case the event is handed over to the mobile device's own functionality for handling the event. However, this must not affect the current WTA user agent context (step 2 in Figure 4.7).

- The executing service has no binding to the event and the context is not protected. The WTA user agent will look for a global binding. Upon a match with a global binding, the old context is terminated and the associated service is initiated (step 3 in Figure 4.7). Should there not be any global binding either, the event is handed over to the mobile device (step 4 in Figure 4.7). In this case the old context could be preserved.

4.9.3 Event parameters

WTA events can bring parameters to the service that is supposed to capture the event. The incoming call indication event, for instance, carries the number of the calling party, if available, to the WTA user agent. The

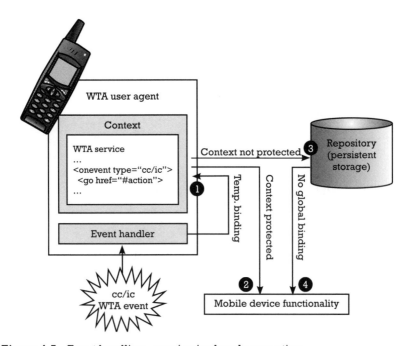

Figure 4.7 Event handling—service is already executing.

service can use the number for making a choice about how to proceed, or just display it to the user.

4.9.4 Example: temporary event binding

For binding an event to an executing service written in WML, the WML **onevent** element can be used. The **onevent** element captures an event of a specified type and binds it to a task. An example:

```
<wml>
 ...
<card id="activecard">
  <onevent type="cc/ic">
    <go href="#notify">
      <setvar name="callerId" value="$2"/>
    </go>
  </onevent>
</card>
  <card id="notify"><p>
    Call from $(callerId)</p>
  </card>
</wml>
```

The card that is currently active defines a binding to the incoming call indication event (cc/ic). The action shown here is that an explicit notification is displayed to the user.

4.9.5 Example: global event binding

If there is no active context in the WTA user agent, an event can be captured by WTA provided there is a channel for that specific event stored in the repository. The binding is realized by giving the channel a name associated with the event to which it is supposed to bind. When the WTA user agent receives a WTA event and finds a matching channel, the first listed resource in that channel will be processed. An example of a channel is:

```
<?xml version="1.0" ?>
<!DOCTYPE channel PUBLIC "-//WAPFORUM//DTD
CHANNEL 1.0//EN" "channel.dtd">
<channel
  maxspace="2048"
    base="http://wap.serviceprovider.com/"
    EventId="wtaev-cc/ic"
    success="success.wml"
```

```
      failure="failure.wml"
  <title>Call Handler</title>
  <abstract>
    This service is handling incoming calls
  </abstract>
  <resource href="firstresource.wml"/>
  <resource href="callhandlingscripts.wmls"/>
  <resource href="voicemail.wmls"/>
</channel>
```

This channel is associated with the incoming call indication event, by assigning the attribute `EventId` with the `wtaev-cc/ic` value. The WTA user agent will start processing the firstresource.wml resource, which in turn uses the other listed resources. The firstresource.wml resource could be a WML deck that presents the user with options for how to handle the incoming call. The callhandlingscripts.wmls resource is probably a set of functions that use WTAI for handling the call. By the name of the voicemail.wmls resource, one would guess that this service also offers forwarding to a voice mail service.

4.10 Building a WTA application

To describe how a WTA service is built using the WAE application framework with WTA components, this section will present an example of what could be called an incoming-call handler service. The example does not claim to be exhaustive in any way but to give an author of WTA service content a sense of what a WTA service could look like.

4.10.1 The incoming-call handler service

Many networks today offer call-forwarding services that can be activated and deactivated as requested by the mobile-device user. For example, the network can be "programmed" to always forward all incoming calls aimed for the mobile device to a specific number, or to forward calls only when the user is busy with another call. The purpose of the incoming-call handler service is to give a user of the WTA-enabled mobile device the possibility to act upon an incoming call at the moment it arrives.

In this example there are four different ways to act upon an incoming call: accept the call, reject it, forward it to a party of the user's choice, or forward the call specifically to the operator's voice mail service. This is what should happen when an incoming call is detected:

- ▸ The incoming call handler service is activated by the WTA user agent, meaning that the user is presented with the options accept, reject, voice mail, and forward.
- ▸ An appropriate action is associated with each option.
- ▸ The specified action is invoked upon the user's selection.

To implement the first step, the event-handling mechanism specified for WTA has to be used. There is a WTA event named *Call Control/Incoming Call Indication* (cc/ic) defined in network-common WTAI. This event should be associated with a channel that is downloaded to the WTA user agent and stored in the repository. The channel must be named **wtaev-cc/ic** to be associated with the incoming call indication (cc/ic) event.

The second step, presenting a list of options to the user on the mobile device's display, is implemented using a WML card. In this example, the associated actions (step three) are also defined in WML cards. All of these cards belong to the same WML deck. The defined actions result in step four: navigation to other cards or invocation of WMLScript functions.

The channel defined in Section 4.9.5 matches the service described in this example. One important assumption is that the service somehow has been granted permission (blanket permission as described in Section 4.3.2.2) to invoke the WTAI functions used in the example.

4.10.1.1   The first resource

The first listed resource in the channel is a WML deck defined in firstresource.wml. This is the resource that will be invoked when the channel is referenced. What happens first is that a card with the list of options is presented on the mobile device's display. The user can scroll through the options and when accepting one of them, the user agent navigates to the card that is associated with the choice. The WML implementation could look like this (because of space considerations, only fragments of the code are presented):

```
<!-- This is a comment: All WML decks must
include this document prologue which is the XML
and document type declaration. -->
<?xml version="1.0"?>
<!DOCTYPE wml PUBLIC  "-//WAPFORUM//DTD WML
1.1//EN"
```

```
       "http://www.wapforum.org/DTD/wml.xml">
<!-- Here is where the code to be executed
starts -->
<wml>
<card>
<!-- This card displays the list of options -->
<!-- First of all the identity of the call,
allocated by the mobile device, and the number
of the caller, both passed with the cc/ic-
event, are captured -->
  <onevent type="onenterforward">
    <refresh>
      <setvar name="id"  value="$1"/>
      <setvar name="callerId"  value="$2"/>
    </refresh>
  </onevent>
  <p>Incoming call:
    <select>
      <option onpick="#accept">Accept</option>
      <option onpick="#reject">Reject</option>
      <option onpick="#voicemail">Voice Mail
      </option>
      <option onpick="#choosedest">Forward
      </option>
    </select>
  </p>
</card>
<card id="accept"
  onenterforward="callhandling-
scripts.wmls#accept(id,0)">
<!-- This card defines the action for accepting
the call -->
  <onevent type="cc/cl">
    <go href="#release">
      <setvar name="id"  value="$1"/>
      <setvar name="cause"  value="$2"/>
    </go>
  </onevent>
  <p>
    Accepting call
  </p>
</card>
<card id="reject">
<!-- This card defines the action for rejecting
the call -->
```

```
  <p>
    ...
  </p>
</card>
<card id="voicemail">
<!-- This card defines the action for forward-
ing the call to the operator's Voice Mail
service -->
  <p>
    ...
  </p>
</card>
<card id="choosedest">
<!-- This card asks the user for the destina-
tion to forward to -->
  <p> Forward to:
    <select>
      <option onpick="#secretary">
       Secretary</option>
      <option onpick="#mother">Mother</option>
      <option onpick="#enterno">Enter Number
      </option>
    </select>
  </p>
</card>
<card id="secretary">
<!-- This card invokes the WMLScript function
that forwards the call, using the number to my
secretary as destination -->
  <onevent type="onenterforward">
    <go href="callhandlingscripts.wmls#
    forward()">
      <setvar name="destnumber"
value="5551000"/>
    </go>
  </onevent>
</card>
<card id="mother">
<!-- This card invokes the WMLScript function
that forwards the call, using the number to my
mother as destination -->
  <p>
    ...
  </p>
</card>
```

```
<card id="enterno">
<!-- This card asks the user for a destination
number and invokes the WMLScript function that
forwards the call -->
  <p>
    Enter number: <input name="destnumber"/>
    <do type="accept">
      <go href="callhandlingscripts.wmls#
        forward()"/>
    </do>
  </p>
</card>
<card id="release"
  onenterforward="callhandling-
scripts.wmls#release(id)">
<!-- This card invokes the WMLScript function
that releases the call -->
  <p> Releasing call </p>
</card>
</wml>
```

The mobile device display could look like Figure 4.8, the view on the left, when the options list is presented.

4.10.1.2 Activating the forward option

Choosing the forward option results in the card with identity **choos-edest** to be displayed, whereby the user is asked to make another choice, the destination for the forwarded call. The view in the middle of Figure 4.8 visualizes the available options as seen from the user. The example might seem somewhat unrealistic since it is not likely that the numbers to forward are coded in WML, nor is it likely that the user is asked to enter the destination number in real time. The solution would rather be to have this list created more or less dynamically from a phonebook or similar. However, this is just an example.

After choosing the "enter number" option the **enterno** card is displayed, which is shown in the view to the right in Figure 4.8. After entering the number, and confirming it (in an implementation dependent manner), a WMLScript function is invoked.

The WMLScript function called for invoking the forward function is defined in a WMLScript compilation unit (a file) that can encompass several functions. Each function in such a file is referenced using hash

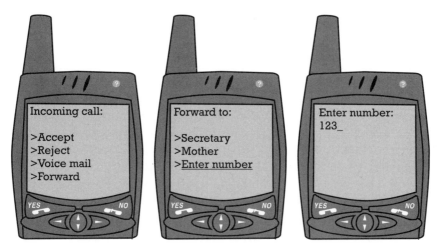

Figure 4.8 An incoming call activates the incoming-call handler service.

marks. An extract from the callhandlingscript.wmls file, with the forward function defined, looks like this.

```
extern function forward()
{
  var id;
  var destnumber;
  var result;
  id = WMLBrowser.getVar("id");
  destnumber = WMLBrowser.getVar("destnumber");
  result = WTAGSM.transfer(id,destnumber)
  if (typeof result != 0)
  {
    Dialogs.alert( "Forward function success-
    fully invoked for "
      + destnumber );
    /* The function call succeeded and the
     result parameter is assigned the identity
     of the forwarded call */
  }
  else
  {
    Dialogs.alert( "The call could not be
    forwarded
      (error code"+ result ")" );
    /* The function call failed for some reason
```

```
and the result parameter is assigned an error
code indicating the cause */
  }
}
```

This forward function invokes the call transfer function defined in the GSM-specific extension [10] to WTAI. The call transfer function needs two input parameters of string type: one representing the identity of the call to be transferred and one representing the destination number to which the incoming call should be transferred. In this example, the cc/ic event provided the identity (id) parameter, and the user entered the destination number (destnumber). Because of mechanisms defined in WML [4], the destnumber parameter is not set in the secretary, mother, or enterno cards, but first after the URI to the forward function has been evaluated and the script is invoked. That is the reason why the forward function does not take any parameters, but retrieves the destnumber and id from within the function. A successful invocation of the call transfer function is indicated by the return of the identity (id) of the transferred call. Should the function fail for some reason, an error code is returned.

The example gives a hint on how error codes can be used. The function return value is first checked against its type to see if it is an integer (WMLScript [5] defines five different types, of which the integer type has code 0). If it is not an integer, we know by the call transfer function definition that the invocation was successful. If it is a negative integer, the return value represents an error code as defined by WTA.

In the example, the user is notified about the result of the function call. This might not be the case in a real service, especially not when it comes to such detailed information as error codes. An author of WTA service content would probably define other actions.

4.10.1.3 Activating the accept option

The "accept" option is interesting since it defines a temporary binding. This is because there will be a subsequent event arriving some time after the incoming call indication event, and that is the call cleared event (cc/cl). The mobile network generates that event when the call is terminated.

The reason why there should be a temporary binding is found in Section 4.9.2. We assume that the context is not protected (this seems to be a valid assumption since the service has not invoked the WTAI function used for context protection). If the service we now have defined does not have a temporary binding to the cc/cl event, it will instead be matched against a global binding. If no such binding exists, the event will be passed to the mobile device's own mechanism for handling such events. Depending on the implementation, this could result in the executing service being terminated. This might not be a problem in this particular service, since the call is about to be released anyway. Still, a solid WTA service would have to control the call during its entire lifetime and not rely on a fallback mechanism offered by the mobile device's core features.

Except for the temporary binding, the implementation of the accept option is similar to that in the forward case. The callhandlingscripts.wmls resource is used since it is assumed to include the functions that invoke the network-common WTAI functions accept call and release call. Note that the second parameter is set to zero in the function call

```
href="callhandlingscripts.wmls#accept(id,0)"
```

This implies that the accepted call will be terminated at the same time the context terminates, even if the calling parties have not decided to do so. A value of one would guard an ongoing call and have it last even if the context ends before the call is brought to an end in a more natural fashion.

References

[1]   Wireless Telephony Application Specification, WAP Forum, Version 16, July 1999.

[2]   Wireless Application Environment Specification, WAP Forum, Version 24, May 1999.

[3]   Wireless Application Environment Overview, WAP Forum, Version 16, June 1999.

[4]   Wireless Markup Language Specification, WAP Forum, Version 16, June 1999.

[5]   WMLScript Language Specification, WAP Forum, Version 17, June 1999.

[6] Wireless Session Protocol Specification, WAP Forum, Version 5, November 1999.

[7] Wireless Datagram Protocol Specification, WAP Forum, Version 5, November 1999.

[8] Wireless Transport Layer Security Specification, WAP Forum, Version 11, February 1999.

[9] Wireless Telephony Application Interface Specification, WAP Forum, Version 31, May 1999.

[10] Wireless Telephony Application Interface Specification, GSM Specific Addendum, WAP Forum, Version 10, February 1999.

[11] Wireless Telephony Application Interface Specification, PDC Specific Addendum, WAP Forum, Version 30, April 1998.

[12] Wireless Telephony Application Interface Specification, IS-136 Specific Addendum, WAP Forum, Version 30, April 1998.

[13] Bray, T., et al., *Extensible Markup Language (XML)*, W3C Recommendation 10, February 1998, REC-xml-19980210, February 10, 1998. http://www.w3.org /TR/REC-xml.

[14] Service Indication Specification, WAP Forum, Version 8, November 1999.

CHAPTER

5

Contents

Integrating WAP Gateways in Wireless Networks

Janet Loughran

5.1   Overview

Providing mobile users with WAP functionality requires mobile networks to become WAP enabled. This chapter will define what this means, and what are the different functions involved. WAP-enabled networks will be able to provide to mobile-network customers all the benefits described in the first chapter of this book.

The capability for mobile networks to provide WAP services to their customers is provided by functionality in the mobile devices themselves and by a network element called a WAP gateway, which provides an interface to the mobile network. Section 5.2 provides details on the location of WAP components in the mobile network.

Section 5.3 lists all functions that a WAP gateway needs to perform, covering those specified by the WAP Forum and therefore standardized, those necessary to integrate the standardized WAP functionality to an actual mobile-network implementation, and value-added services provided by manufacturers.

Section 5.4 gives some ideas on future enhancements for the WAP gateway. Finally, Section 5.5 lists some features that WAP gateway manufacturers are using to differentiate their products from those of their competitors.

5.2   Positioning of WAP functionality in a mobile network

The following lists the WAP components:

- WAP gateway;

- WAP-enabled mobile devices;

- WAP origin server;

- Wireless telephony application server.

5.2.1   WAP gateway

The primary function of the WAP gateway is to provide a link between a mobile network and the Internet, so that WAP-enabled mobile devices can request WAP services and information from Web servers. In this way, the WAP gateway allows WAP-enabled mobile stations to access applications hosted on standard origin servers, the servers that host and deliver the WAP content, and are usually based in the mobile operator's network, as well as from other places (e.g., service provider's premises, content provider, service broker, the Internet, etc.). See Figure 5.1.

WAP services will be developed on the origin servers using WML and the WMLScript. See Chapter 2 for more information on developing WAP applications and services. WML is based on XML, while WMLScript is based on ECMAScript (a standard from the European Computer Manufacturers Association). The WAP gateway encodes the WML or compiles the WMLScript into compact binary form, and forwards the resulting data to the mobile device. These encoding and compiling functions ensure efficiency of transmission across the limited bandwidth offered by the mobile air interface.

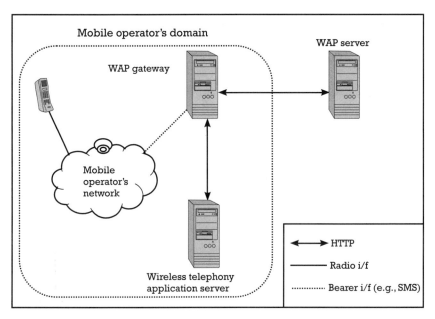

Figure 5.1 The high-level WAP architecture and its relationship to the mobile operator's network.

5.2.2 WAP-enabled mobile devices

WAP-enabled mobile devices will offer to the mobile user the ability to retrieve WAP content from the origin server via the mobile network and the ability to render that content in an appropriate understandable format on the mobile device display. The user selecting a WAP service from the man-machine interface (MMI) of the mobile device results in a URI being sent to the WAP gateway as the WAP service request. The return of content from the origin server to the WAP gateway is based on the standard URI retrieval mechanism. This content is then subsequently sent to the mobile device to complete the service transaction.

5.2.3 WAP origin server

A WAP server is a standard origin server that provides content in WML or WMLScript formats, rather than HTML. Resulting from receipt of a URI request from the WAP gateway, the requested content is returned to the WAP gateway, in the same way HTML content would be returned.

5.2.4   Wireless telephony application server

It is anticipated that operators will use custom-developed WAP services to differentiate themselves from competitors. This can be achieved by offering subscribers advanced telephony and messaging services (e.g., advanced call control and messaging access). These services are likely to be hosted on specially enabled WAP servers located within the operator's network. WTA servers will access the operator's networks via interfaces such as signaling system 7 (SS7) (as well as MAP, ISUP (ISDN user part), CAP, and INAP (intelligent network application part)).

The WTA will contain a repository management application, which will allow the operator to alter and update the repository content on the WAP handset using HTTP/push as illustrated in Figure 5.2. Typically, a WTA service is required to react to particular network events. For example, a WTA service may present the user with various menu options in response to an incoming call event (i.e., accept the call, forward the call, or divert to voice mail). Because of the inherent latency associated with the current WAP bearers, it is essential that this menu is located locally on the phone instead of being downloaded to the phone via a URL request each time this particular network event occurs. This network event table showing incoming call versus menu options is known as the repository content. Each WTA service will require its own repository content that can be administered from the repository management function. Other examples of application areas for WTA services are: advanced messaging, directory services, and pre-call answer menu. Chapter 4 goes into more detail describing WTA.

5.2.5   Additional support offerings by WAP gateway manufacturers

In addition to the WAP components specified previously, parties interested in promoting the WAP technology are likely to offer a range of other services, examples of which are listed.

5.2.5.1   WAP developer's programs

All parties involved with, or having interest in, the deployment of WAP-enabled networks will be keen to promote the creation of new WAP and WTA services. They will wish to ensure that the benefits that WAP can offer to mobile users are explored to their full potential, especially in the early stages of system deployment. WAP service providers want to explore the target deployment areas for the technology and get feedback from areas of the market that will utilize their offerings.

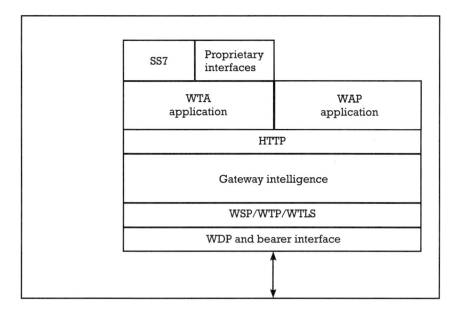

Figure 5.2 Positioning of WAP and WTA applications within the WAP gateway protocol stack.

There will be opportunity with WAP for product differentiation between the offerings of different mobile network operators. As WAP components have become commercially available, the interest in development of WAP content and value-added services (VAS) has increased dramatically. Another area ripe for take-up by a WAP developer's program will be with access to wireless telephony application servers, enabling developers to create WTA with which users can interact, making use of both telephony and data services.

Manufacturers providing their products in this way enable both promotion of their products and provision of an arena for industrial alliance negotiations. However, these practices not only promote the products being offered, but also give an awareness of WAP within the telecommunications and datacommunications fields. For the sake of the technology, many participants need to survive. What is the good in having an open standard with the participation of only a handful of industry participants, be it mobile network operators, infrastructure providers, vertical market vendors, or Web content providers? At this early stage in the WAP game, everyone will be pulling together.

With such developer programs, application and content providers will be able to develop, test, and deploy WAP applications that interface with WAP components in a real-time environment.

5.2.5.2   Provision of WAP applications

As part of the drive for uptake of the WAP product set, WAP gateway manufacturers have developed trial applications, which can be hosted on the gateway platform or on a separate platform. These are then offered as part of trial evaluation kits to interested parties, allowing early visibility of the potential of WAP services.

5.2.5.3   HTML content availability

Manufacturers may optionally provide tailored HTML to WML filtration capability as part of the WAP gateway. This can allow access to new or existing HTML-based applications from WAP-enabled devices.

5.2.5.4   Vertical market relationships

Manufacturers will tend to develop relationships with service providers in vertical markets, with a view to providing a broad range of services on WAP gateway platforms. These will include but are not limited to:

‣ Unified messaging (see Chapter 10);

‣ Corporate work flow and database access;

‣ Directory and information services;

‣ Banking services (see Chapter 11).

5.2.5.5   Vendor partnerships

In order to supply full WAP solutions, a wide range of diverse telecommunications and software development technologies are required. Manufacturers are likely to rely on their strength areas of expertise and form partnerships with other vendors to ensure end-to-end provision of WAP services. This is a very sensible approach for vendors to take. For example, although the worlds of telecommunications and datacommunications have been shown to be merging for a lot of application areas, the basis of the technologies themselves are so diverse that experts in one of these fields are likely to be novices in the other.

5.2.5.6 Early network integrated test environments

To provide early multivendor integration possibilities, WAP component manufacturers will be interested in being involved with the provision of end-to-end multivendor WAP systems for early network integration test (NIT) environments. This type of coordination between vendors is common practice with the development of components for open systems.

5.2.5.7 WAP evaluation kit

WAP offers services to mobile users that for the most part have not been available with other technologies or systems to date. The responsibility of offering entirely new types of services is that the users need to be educated as to what the new opportunities are, and how to make the best use of what will be available to them. This will not be a one-way education either, as the industry will have a lot to learn by listening to the feedback provided by new users. This method of gathering requirements by examining the market's needs is referred to as task analysis and will provide valuable information to the WAP industry on how to proceed with WAP. It will also gain the interest of potential users at an early stage.

Thus, WAP component manufacturers will often provide evaluation kits for such reasons and also assist potential customers to:

› Understand the architecture and dynamics of WAP system infrastructure;

› Understand how to develop and implement WAP applications;

› Access demonstration WAP applications hosted on the WAP origin servers;

› Begin to evaluate and understand WAP products and services;

› Develop and test some initial WAP applications.

5.3 Functional requirements of a WAP gateway

The WAP gateway is an enabling platform that will provide mobile operators with the ability to introduce WAP-based services into their networks. The requirements in terms of functionality of the WAP gateway can be divided into three areas that will be discussed in the next three subsections:

▶ Section 5.3.1—Standardized functionality specified by the WAP Forum;

▶ Section 5.3.2—Functionality required to integrate the standardized WAP functionality with a customers' network;

▶ Section 5.3.3—Value-added services provided by manufacturers.

5.3.1 Standardized functionality specified by the WAP Forum

The WAP stack implements the WAP protocol and allows the transport of content between the WAP gateway and a WAP-enabled handset. WAP protocol stacks are required on both the handset and the WAP gateway so that peer-to-peer protocol connections can be performed. The standardized WAP stack consists of the following layers (see Figure 5.3):

▶ Wireless datagram protocol layer;

▶ Wireless transaction layer security;

▶ Wireless transaction protocol layer;

▶ Wireless session protocol layer.

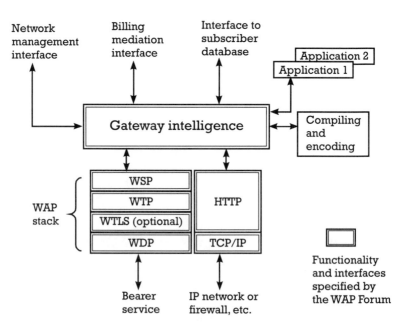

Figure 5.3   High-level architecture of WAP gateway and interfaces.

5.3.1.1 Wireless datagram protocol layer

The communication mechanism actually used to transport data between the WAP gateway and the handset is referred to as a bearer. Typically, bearers would be short message service (SMS), circuit-switched data (CSD), unstructured supplementary service data (USSD), cell broadcast (CB), and general packet radio service (GPRS). Different mobile bearers exhibit very different bandwidth and latency characteristics (e.g., GSM SMS messages are limited to 160 characters).

The WDP layer performs all necessary bearer adaptation (i.e., adapting the data for either transmission to the mobile network, or following receipt from the chosen bearer, sending them to the next layer in the WAP protocol stack). In general, adaptation involves breaking up the data into fragments of an appropriate size for the bearer, and interfacing with the bearer network to transport the data. For example, GSM SMS adaptation involves fragmenting the data into segments of 140 octets and sending this data in short messages (SM) to the handset. The WDP layer on the handset reconstructs the data from the received SMS and presents them to the higher layers of the WAP stack.

Since adaptation is performed by the WDP layer, the higher layers of the WAP stack do not need any knowledge of the bearer. This allows the higher layers of the WAP stack and the applications and browsers to remain independent of both the mobile network and the bearer.

5.3.1.2 Wireless transaction layer security

The WTLS layer provides the security layer for the WAP stack by providing privacy, data integrity, and authentication between two communicating applications. Data are compressed and encrypted before being sent over WDP, and are decompressed and decrypted when received from WDP. The WTLS protocol layer is an optional feature specified in the WAP 1.1 standards, released in June 1999. However, many WAP handset manufacturers are unlikely to support WTLS in the short term; therefore, gateway manufacturers may decide not to include this in their early WAP gateway releases to the market. Refer to Chapter 7 for a more detailed discussion on WTLS.

5.3.1.3 Wireless transaction protocol layer

WTP is a lightweight transaction-oriented protocol designed to run on top of a datagram service (i.e., WDP). It provides retransmission and acknowledgment services, relieving the upper layers of these tasks. Together with WDP, it forms the transport layer of the OSI seven-layer communication stack model. The service user of the WTP layer (i.e., the wireless session protocol layer) would choose to use this protocol for greater integrity of the data being sent.

5.3.1.4 Wireless session protocol layer

The WSP layer provides session services to the WAP application layer allowing the exchange of information within a session, and also is a service user to the WTP layer. WSP provides two services:

1. The connection-orientated service allows a session to be reliable by using the acknowledgment and retransmission and facilities of the WTP layer. Also, with the connection-orientated mode, the mobile device and the WAP gateway can negotiate a mutually acceptable set of capabilities (e.g., maximum send data unit (SDU) size). The service also allows the session to be suspended and resumed on another bearer if required.

2. The connectionless mode service provides a service without acknowledgments or retransmissions between the client and the WAP gateway. Messages utilizing this type of service do not make use of the WTP layer services, but instead are passed straight to the WDP layer.

5.3.1.5 Compiler and encoder functionality

This functionality is specified by the WAP Forum and forms part of the presentation layer requirements of the WAP protocol stack. The content information that is sent from the WAP origin server to the WAP gateway will either be in WML or WMLScript format. The encoder encodes the WML into compact binary form, and the compiler does the same for WMLScript, to enable efficient transmission to the mobile device.

5.3.2 Functionality required in integrating the standardized WAP functionality to an actual mobile network implementation

Standards bodies like the WAP Forum, ETSI, etc., specify the peer-to-peer relationships between entities in network components in order that open systems can be produced. They specify the protocols of layers and what a service-providing layer offers to the layer to which it provides service. For example, the four specified layers of the WAP protocol stack will be defined in terms of "what they must do." The "how they do it" remains in the realm of the product manufacturers. The decisions by manufacturers of how to implement their products will lead to differentiation and added value among offerings.

There will be many choices on how to implement the protocol stacks, based upon such requirements as: performance, distribution, capacity, upgrading, and cost, to name a few. Consideration of these requirements will determine the design criteria to be met by implemented products.

This section contains descriptions of the basic functionality needed to support the requirements in the WAP specifications, in order for products to be implemented, and to integrate the WAP gateway functionality with existing mobile networks.

"Gateway intelligence" is a term used to describe the functionality that needs to exist in order to implement a commercial gateway product. The WAP Forum has specified the components to allow peer-to-peer communications between a mobile device, a WAP gateway, and the Internet. What is required in addition to support a commercial implementation is listed here:

‣ Gateway management function;

‣ Gateway intelligence interfacing function;

‣ Configuration data.

The following list details possible value-added functions:

‣ Scalability, flexibility, and distribution;

‣ Event managing function;

‣ Gateway management function;

‣ Push applications;

‣ Billing data interface;

‣ Subscriber data;

‣ Caching of wireless content.

These concepts are described here.

5.3.2.1 Gateway management function

A management function will be required for management of all key components of the WAP gateway. How it will be implemented is entirely an implementation decision among manufacturers, and the method of implementing can offer a lot of added value. This concept is developed further in Section 5.3.3.

5.3.2.2 Gateway intelligence interfacing function

A function is required for interfacing between the WSP, compiler and encoder, Internet and intranet, push application's API, and whatever other gateway intelligence functions may be provided.

The WAP specifications mandate the use of HTTP 1.1. URI requests are accepted from the WSP and passed to the HTTP client, which will retrieve the associated WAP content from an appropriate origin server using HTTP protocol over TCP/IP (i.e., the standard Internet protocols). If the request is serviceable, the origin server responds with the requested content.

Content types defined by WAP have a compact binary format suitable for efficient over-the-air transmission to the mobile devices. If the response from the origin server is text WML, it is passed to the encoder for conversion to bytecode (binary format), and if the response body from the origin server is text WMLScript, the compiler performs the conversion to bytecode. In addition, the standard text HTTP headers have an equivalent compact binary format defined by WAP. These are also passed for conversion to bytecode.

Additional presentation layer functionality transcodes the content provider's character set to the mobile client's preferred character set. Content other than WML or WMLScript will pass through without change. The resulting content is subsequently passed to WSP for transmission to the client.

5.3.2.3 Configuration data

To provide flexible gateway configuration, gateway intelligence will manage the configuration data. The parameters contained will be held in some means of persistent storage and read on system start-up and also when signaled to do so by the gateway intelligence function. It is an attractive feature for a gateway to be able to dynamically reconfigure its operation without disturbing ongoing sessions. In fact, mobile operators will most likely gauge this capability as being a requirement. It is important that a WAP gateway can be as flexible as possible in managing its configuration data.

5.3.3 Value-added services provided by manufacturers

This section details extra functionality that could be provided by gateway manufacturers to differentiate their products and thereby provide additional value. Obviously this functionality is expected to grow as both the WAP standards and the products develop, but this is a snapshot of what is available at the time of writing this chapter.

5.3.3.1 Scalability, flexibility, and distribution

Modern software design methods enable system design with built-in upgrade capability. As deployment of WAP services grows, scalability of the WAP gateway will become a necessity. Gateway design should allow for increasing functionality without requiring a complete architectural redesign.

An example of functionality to be introduced is that for handling new bearers in the WDP layer: packet data services will become widespread as the industry moves steadily towards the advent of high-bandwidth third-generation mobile systems.

The WAP specifications have been defined as a set of protocol layers, with defined interfaces between each layer. These layers should be designed and implemented in such a way as to ensure low coupling, high cohesion, and abstraction of the interfaces between layers. Age-old design principles these may be, but they are the key ingredients to providing scalability of design.

The service access points (SAP) interface definitions in the WAP specifications ensure that the information flowing between layers is open. The SAP provides a well-known interface that other components of the software and applications use to establish communications with the layer. If it is decided to use industry standard communication protocols for the interlayer communications, gateway layers can easily be distributed, the interlayer communications can use transport provided within the mobile network, and manufacturers can avail of off-the-shelf standard interfacing software.

Depending on the configuration of a mobile network, and also on the requirements of WAP applications, it may be desirable to distribute the WAP gateway over more than one geographical location. Standardized interfaces between gateway components would enhance suitability of a gateway for distribution.

5.3.3.2 Event managing function

As a constituent part of a real-time service network, the WAP gateway will most likely be required to provide details of its run-time operation to a gateway controlling function. One method is to log occurrences of events as they happen in areas of gateway operation. In this way, a gateway managing function can monitor operation and can trigger action if, for example, an undesirable event occurs. The classification of an event could be decided by this managing function, and a particular event may be reclassified with a change to the managing component only.

An event managing function could collect all events recorded by the WAP stack (e.g., SMS received, URI decoded, origin server access refused). Example types of the event may be:

- Billing events;
- Information events;
- Alarm events.

5.3.3.3   Gateway management function

As defined in the previous section, this functionality will provide added value to the gateway, and so listed here are some typical areas where this could be realized. The gateway management function would provide an open interface to a network management GUI, allowing the operator to request management operations.

Typical functionality could be:

- Start-up/shutdown of WAP gateway;
- Start-up/shutdown of individual components;
- Monitoring/restart of components;
- Continual monitoring of all of the components for which it is responsible;
- Automatic restart of a failed component;
- Access to/update of configuration data;
- Ability for the operator to manage the configuration data for the WAP gateway;
- Management of ongoing WAP sessions within the gateway;
- Output of statistics.

It is desirable for a gateway to be able to gather statistics to assist with capacity planning, for example, to monitor peak-load situations. This could be achieved by components being signaled at regular intervals to output their counters to a database. The statistics could be accumulated from the database at regular intervals to form a statistical view of the traffic within the system.

5.3.3.4 Monitoring of critical alarms

By addressing this aspect of network integration, the operator could be provided with sophisticated alarm handling, ideally with the ability to track alarm histories.

The WAP gateway is only one element of the operator's network, which will also contain other elements such as short message service center (SMSC), home location register (HLR), routers, switches, etc. It is a requirement that the gateway should provide an open interface to the operator's network management system (NMS) using simple network management protocol (SNMP). An NMS allows the operators to manage all of the elements in their network in an integrated manner and from a single platform. The SNMP interface will allow the WAP gateway to be managed from the operator's NMS. Any faults experienced on the WAP gateway will be alarmed automatically to the NMS where they can be dealt with centrally.

5.3.3.5 Push applications

As well as providing the familiar request/response transaction support, WAP also defines a push transaction with which an application may send unsolicited information to a client. At present, the WAP Forum has not defined an end-to-end architecture for push, or what particular push functions the gateway should or must provide. In the absence of a standardized approach, gateway manufacturers may choose to provide proprietary push application interfaces. However, once the standards are complete, then WAP gateway manufacturers will have to ensure that their products conform to them.

5.3.3.6 Billing data interface

The fundamental reason for an operator to introduce a new service is to increase revenue. Therefore, the operator must be able to bill subscribers for use of that service. In order to provide commercial WAP services, it is essential that the gateway provides interfaces to the customer's subscriber database and billing system. It is still not clear how operators will bill for WAP services, and therefore it is desirable that any solution provided is flexible and can easily be adapted when billing becomes more mature.

A simple provisioning model that can be assumed is that WAP sub-scribers are provisioned from a central location within the customer's network (e.g., the information may be copied from data stored in the net-work's HLR), and then this will be stored in a database. The gateway needs to be able to access this data when necessary through an external, ideally open-standard interface.

The WAP gateway can be designed to gather extensive billing data for each transaction (e.g., download of content made by a subscriber, URIs visited, time taken for download of content, etc.). Flexibility in changing what might be deemed as billing data is desirable. This billing data could be stored within the WAP gateway and made available to the operator's billing system.

To facilitate interaction with disparate billing systems, it is required that billing data be stored in the WAP gateway in a generic and flexible format. This flexibility adds enormous value to the WAP gateway by allowing operators to introduce and bill for new services easily without having to make changes to their existing billing systems. Figure 5.4 shows network integration of the WAP gateway with a customer billing system.

5.3.3.7 Subscriber data

The gateway needs to provide an interface to subscriber data in order to deploy the service into a commercial network. The WAP gateway can provide basic authentication as to whether a subscriber has access to

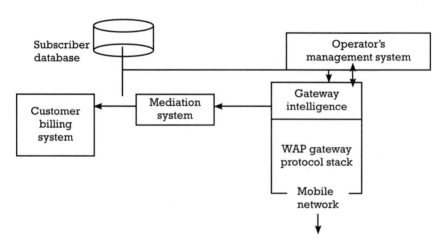

Figure 5.4 Billing interface to the WAP gateway.

WAP services. Provision of this functionality and the type of information being requested from the database is likely to vary among WAP gateway customers.

Typical examples of subscriber data access could be:

▸ To check whether a subscriber has access to WAP services in general, or to a particular URI, or to a specific application. This could be achieved by sending the subscriber's mobile station integrated international service digital network (MSISDN), obtained from the bearer, to the customer subscriber database to verify that the subscriber in question has been provisioned for WAP services.

▸ To determine the bearers to which the subscriber has subscribed (e.g., SMS, CSD), and to determine the subscriber's preferred bearers for particular services or applications.

▸ Blacklisting, spamming—service management issues.

A function of gateway intelligence may be to manage the set of accessible URIs. The actual range of URIs that the subscriber will be allowed to access on origin servers might be dictated by a URI white list function. It could be possible to provide the same URI white list to all WAP subscribers.

5.3.3.8 Caching of wireless content

In the same way as Internet browsers cache information for easy and quick re-retrieval, the gateway intelligence functionality can accommodate caching, thereby reducing processing and response times for the requesting mobile-device user. Obviously this is a very attractive offering for mobile operators, as this reduces the time that radio resources are tied up for any WAP session, especially desirable when we are dealing with circuit-switched data bearers.

5.4 WAP gateway future enhancements

Additional features to be included in the WAP architecture are currently being discussed within the WAP Forum. Some of these possibilities are listed here.

5.4.1 Push applications

We refer to Chapter 6 for a detailed description of the implementation of push services with the WAP framework.

5.4.2 Security

Many WAP-based applications will require secure transactions to be supported (e.g., personal banking). WTLS will secure WAP communications between the mobile device and the WAP gateway. Support for security on the gateway to origin server interface will be provided by use of HTTPS, which is a secure version of HTTP. Implementation of both these security mechanisms will provide true end-to-end security provision for WAP services.

5.4.3 Provisioning server

WAP gateway manufacturers may offer dedicated WAP provisioning servers to customers to enable management of access to WAP services. These will contain all aspects of the subscriber database as previously discussed, plus all functionality surrounding managing subscriber access. The provisioning server would provide: persistent storage of subscriber data, a provisioning interface which can handle bulk provisioning, and a suitable directory access protocol towards the WAP gateway.

5.4.4 New generation mobile networks

The mobile networks that will offer WAP services are continually evolving, and therefore so must the support offered to them by any service functions such as WAP.

Manufacturers of WAP gateways must design their products so that they are future proofed. This should be in terms of how the capabilities of WAP itself will develop, and also to address the needs of up-and-coming mobile infrastructure characteristics. New mobile network air interfaces will require new types of telecommunication bearers, like GPRS, EDGE (enhanced data rates for GSM evolution), and W-CDMA (wideband code division multiple access). New transport mechanisms and architectures within these networks might affect how WAP components interface to them.

With the increased bandwidth, for example, W-CDMA systems with mobile users being able to achieve up to 384 Kbps, WAP will have a lot to offer its users.

5.4.5 Interim proprietary solutions

Until specification of the WAP gateway functionality by the WAP Forum is complete, manufacturers in some cases will offer proprietary solutions. However, forward-thinking manufacturers will ensure that their product architectures are flexible, and that they will be able to adapt as easily as possible to additional protocols and functional specifications to be made by the WAP Forum.

5.5 The WAP gateway—product differentiation factors

WAP gateway manufacturers will use various techniques in order to differentiate their products from their competitors, with the hope of gaining market advantage. The following list summarizes some of the features addressed.

‣ Competitive entry-level pricing;

‣ Value-added service provision for network operators;

‣ Products that aim to be open, scaleable, modular, and configurable, ready for commercial deployment with billing and subscriber database interfaces;

‣ Main customer target choices: large corporations, banking institutions, Internet service providers, mobile e-commerce, vertical markets (e.g., fleet management and wireless messaging and information companies);

‣ Business tool offerings: encryption, billing, directory services, system management, subscriber database provisioning;

‣ Full-billing solutions on the WAP gateway platform;

‣ Billing interfaces to proprietary network management systems;

‣ WAP gateway to application server integration.

CHAPTER

6

Contents

Introduction to WAP Push Services

Bo Larsson

6.1   Introduction

When Tim Berners-Lee invented the World Wide Web (WWW) in 1990, it was designed to provide scientists at the CERN Laboratory in Switzerland with the ability to structure their documents and allow other people to reference their work in a convenient and efficient manner using a shared network for publication—the Internet. The means to provide this functionality—HTML, HTTP, and URLs—were not designed to be versatile tools for creating what we today call Web-based services, even though they have turned out to be useful as such.

Both HTML and HTTP have been revised several times to meet the needs of the broadened use of the Web. However, the fundamental idea of how to access content on the Web using HTTP has not changed. The user types in a URL, or uses options like bookmarks that identify the information he or she wants to access, and then pulls the

117

information from a server on the Internet. The opposite way, allowing a server to push information to a user, is limited as of today. This is a flaw, since many services would benefit from that ability.

Consider, for example, weather forecasts. Today I have to visit the weather bureau's home page potentially several times a day to find out if there is an updated forecast or possibly even a storm warning. With push I could have it delivered to me directly when it is updated, saving me the inconvenience of having to repeatedly look for it. There is an abundance of similar use cases. A hackneyed but very intelligible example is stock quotes, where timely access to information is of the uttermost importance for investors.

The WAP Forum acknowledged the need for push in mid-1998 by creating a drafting committee dedicated to designing a framework for push in WAP, a work that was completed a year later and subsequently included in WAP 1.2, released in December 1999. The framework addresses not only the wireless sphere, but also the Internet domain. In other words, it is an end-to-end solution.

Before we enter the push framework, here is a definition of push and a quick insight into existing technology.

6.2   Definition of WAP push

Throughout this chapter the following definition of WAP push applies:

> "The ability to deliver arbitrary content between a push initiator and a specific user agent on a mobile client in an asynchronous manner. Push initiators may reside on Internet servers or on dedicated WAP servers."

As we shall see later in this chapter, WAP servers communicate with the mobile client directly, while Internet servers communicate with clients via an entity called push proxy gateway.

6.3   What do we have today?

As mentioned in the introduction, there is no such thing as a "genuine" HTTP push on the Web today; it is always the client that initiates any form of communication. If we consider other applications on the

Internet, e-mail is a good example of an application that makes asynchronous delivery of content possible. But let us stay on the Web for a while.

6.3.1 Push on the Web

It was mentioned that HTTP does not support push of content from an Internet server to a client. But browser manufacturers claim to have support for push in their products. How does this fit together? The key to the riddle is scheduled pull—a solution that relies on the browser's ability to check for new or updated content at regular intervals—something the user will experience as push. Let's take a little closer look at one of the most widespread solutions, Microsoft's Active Channels.

When Microsoft introduced Internet Explorer 4, the big news was its active channels. These channels allow users to subscribe to information, which is presented to them using desktop items, HTML formatted e-mail, or a special screen saver with the ability to render Web content.

Active channels are implemented using a technology called channel definition format (CDF). CDF is an application of the XML—the metalanguage that has been evangelized by the Internet community in recent times—allowing content providers to specify the behavior of their channels. The CDF content, which is stored on the user's computer when he or she subscribes to the channel, essentially contains information about what content should be downloaded to the user's computer, and how often the user's client should check for updates.

Let's illustrate how this could work in real life by revisiting the weather bureau example given in the introductory section. Assuming that the bureau's active channel is not present in Explorer's preinstalled list of channels, the user can either visit the channel guide (a guide offered by Microsoft where service providers can expose their services), or visit the bureau's Web site in order to subscribe to the channel by clicking an icon. Once subscribed, the user's client will automatically download the latest weather forecast and display it to the user. Let's assume that the weather bureau updates its forecast twice per day, once at noon and once at midnight. Then the bureau would design its channel so the user's client automatically—without user intervention—downloads a new weather forecast at those times. If the bureau updates its forecast in a more unpredictable manner, it may choose to instruct the client to look for an updated forecast once an hour instead. If no new forecast is available when the client looks for one, nothing will happen.

6.3.2 Push in the wireless domain

Since the Internet has not yet entered the wireless domain to any large extent, we do not find any solutions there that are similar to the ones used on the Internet. Instead, network-originated delivery of information is accomplished using means provided by the network itself, for example, SMS. The services that can be delivered using such means are, however, limited to be fairly simple since only text can be presented to the user. There is, for instance, no standardized way of providing the user with navigational mechanisms that allow him or her to use the received SMS as a starting point for a service. For example, if the received SMS contains news headlines, the user cannot request a certain news item to be displayed.

However, despite its limitations, SMS services have turned out to be a very useful tool in delivering information to mobile users. Popular examples include voice mail notifications, sport results, jokes, and other kinds of information where a small text message is sufficient.

6.3.3 Can the solutions converge?

Now that WAP enables access to Internet content in a way similar to what we are used to from our desktop computers, can we simply adopt the push technologies used on the Internet today to enable push-alike functionality in the wireless domain as well? Well, it would work in theory, but not in reality.

One of the key differences between wired and wireless networks is the bandwidth they offer. In wireless networks, bandwidth is most often a scarce resource, which implies not only lowered performance, but also higher costs to the user. If, for example, Microsoft's Active Channels were to be adopted by WAP, the overhead traffic it causes would simply be too big to align with the requirements on efficient bearer utilization in wireless networks. Let's once again illustrate with some examples.

As we learned from the weather bureau example, the more unpredictable the information to be delivered to the client changes, the more frequently the client needs to check for updates using CDF. This would not be the case if "true push" were available, since the bureau then could deliver a forecast whenever it is updated without prompting from the client. Then consider some of the services that would benefit from using push in WAP, for example, messaging services (e-mail, fax, voice mail, etc.; see Chapter 9 for more details on WAP and unified messaging), the classic stock monitoring service (see Chapter 10 on mobile financial services), and telephony services. They all have a couple of things in common—they are triggered in a nondeterministic manner, and the user experience would to various extents be considerably deteriorated if the event that triggers the service is not served in a timely manner. Personally I would not find it acceptable to wait more than a couple of minutes before I am notified about a new e-mail, and the stock quotes monitoring service may be even more time-sensitive if one wishes to strike while the iron is hot. Finally, many telephony services require real-time handling, implying that the need for polling a server for new information could make the service impossible to implement. An example of such a service is call screening—a service that notifies the calling party that the called party is busy and gives him or her some options (leave a voice message, send an e-mail, be forwarded to the secretary, etc.).

These examples illustrate that a model based on scheduled pull, such as Microsoft's CDF, would not work in the wireless domain. Considering the type and amount of wireless push services we can expect to be deployed, the number of messages sent over the air would simply cause an undue load on the network, especially if low-bandwidth bearers like SMS are used. Another fact to bear in mind is the positive correlation between bearer utilization and the mobile device's power consumption. Increased network traffic would thus imply lowered standby time for devices since their batteries would be emptied faster.

WAP has instead extended the Web model by introducing push capabilities in the WAP protocols so that true asynchronous delivery of content can be accomplished. The next section will describe how this is accomplished.

6.4 The WAP push framework

As said in the introduction, WAP defines an end-to-end solution for push; that is, delivery mechanisms to be used both on the Internet and in the wireless domain are defined. Before we go into details on how this is achieved, it is important to have a clear understanding about some terms used.

6.4.1 Gateways, proxies, and servers

The WAP model relies on a client/server paradigm where an intermediate WAP gateway connects mobile clients and Internet servers with each other. In the case of pull, the WAP gateway essentially listens to requests from the mobile client, retrieves content from an Internet server, and usually performs transformations on the content before it is returned to the mobile client. Since the WAP gateway not only translates between WAP and Internet protocols, but also can perform other operations (e.g., transform WML into its binary form), it also contains proxy functionality despite its name.

When push enters the WAP architecture, the functionality is extended with capabilities that allow push initiators (see Section 6.4.2) on Internet servers to deliver content to mobile clients without any explicit request from them. Once the user of the mobile client has sub-scribed to a service, it is automatically delivered to him or her. This places new requirements on the WAP gateway. In order to distinguish between pull and push functionality, the push specifications use different terms to refer to the entities providing these functions: the method proxy gateway (pull) and the push proxy gateway (push). While it is obvious how the push proxy gateway received its name, the method proxy gateway received its name due to the fact that it listens to method invocations from the mobile client [1]. Figure 6.1 illustrates the architecture.

Hence, a WAP gateway may contain both a method proxy gateway and a push proxy gateway, or one or the other. While many networks benefit from having both implemented, there are certain cases when only

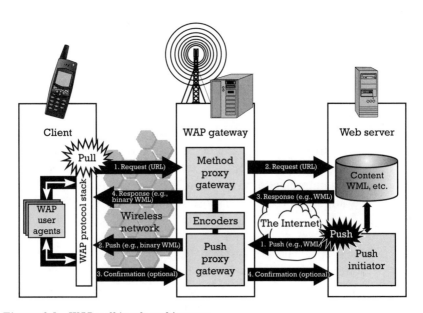

Figure 6.1   WAP pull/push architecture.

one of them is needed. For example, in a one-way paging network it would make sense to only have a push proxy gateway.

There are also cases when the WAP gateway is not needed at all. This is the scenario when the mobile client communicates with a dedicated WAP server directly, using pull and/or push technology. A WAP server is a server with a WAP protocol stack and bearer interface, enabling it to communicate with mobile clients without involving a WAP gateway. The most compelling use for stand-alone WAP servers is when one wishes to obtain end-to-end security using the WTLS protocol (see Section 6.5.3 and Chapter 7).

6.4.2 Push initiators

In the previous section, the term push initiator was introduced. A push initiator is an application on a server that takes the initiative to push something to the mobile client. How the application is triggered depends on the service. For example, it might monitor the price of a stock in order to initiate a push informing the user that a certain condition is satisfied, or it might communicate with a voice mail server to notify the user when there are new voice mails. Push initiators may reside on Internet servers or on dedicated WAP servers as shown in Figure 6.2.

Push initiators on ordinary Internet servers communicate with the push proxy gateway using the push access protocol, which in turn communicates with the mobile client using the push OTA (over the air) protocol. Push initiators on WAP servers use the OTA protocol directly. The

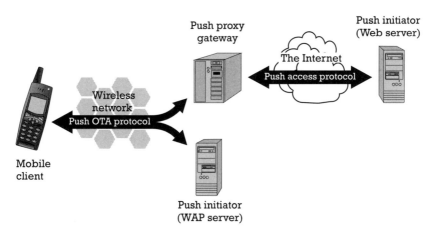

Figure 6.2 Push initiators.

subsequent sections will describe the push access protocol, the push OTA protocol, the functionality provided by the push proxy gateway, and, finally, the behavior of the mobile client.

6.4.3 Push access protocol

The push access protocol [2] provides push initiators on the Internet with a means to communicate with the push proxy gateway, as depicted in Figure 6.2. The protocol is used to submit content that should be delivered from the push initiator to the mobile client and to perform various other push-related operations. This protocol is the part of the push framework that will be exposed to "the masses" (i.e., all the push initiators (service providers) that will use it). The other parts of the framework mainly concern equipment implementers.

6.4.3.1 Protocol design

When communicating with the push proxy gateway, the push initiator needs to be able to send content (WML, WMLScript, bitmaps, etc.) to the mobile client and push-related control information (recipient address, preferred bearer, delivery time constraints, etc.) to be used by the push proxy gateway. The push proxy gateway also needs to be able to send information to the push initiator, for example, to inform it about the final outcome of a push submission. All operations in the push access protocol are based on a request/response model; that is, for every message sent in one direction, there is a corresponding message in the other direction.

Using Internet standards, the push access protocol is designed to be independent of the underlying transport protocol. The push-related control information sent between the push initiator and the push proxy gateway is sent as a clear text XML document. When a push initiator makes a push submission, this XML entity is bundled with the content to be delivered to the mobile client in a multipart. If you are not familiar with multipart, think of it as a container that can hold a number of content items, but is treated as a single item.

So, to keep it simple, the underlying transport only needs to be able to deliver a single piece of content between the push initiator and the push proxy gateway, and vice versa. The current specification defines how this is accomplished over HTTP, while it is most likely that other protocols will be adopted in the future. A sure candidate is the simple mail transfer protocol (SMTP, used for sending e-mail), which would better accommodate transient push initiators (those who are not necessarily always on-line). The reason for choosing HTTP as the initially supported protocol is simple—almost every server on the Internet supports it, ensuring a high adoption rate. The usage of HTTP is illustrated next.

6.4.3.2 Protocol features

The most fundamental operation provided by the protocol is the push submission (i.e., the operation used by a push initiator to instruct the push proxy gateway to push some content to a mobile client in a particular manner). Figure 6.3 illustrates how HTTP is used to accomplish this.

The push initiator uses an HTTP request method called POST to send the push message to the push proxy gateway. In contrast to the HTTP request method GET, which a browser uses when it loads a Web page, it is possible to also include content in the POST method. This content is the multipart mentioned previously. When the push proxy gateway receives the post, it decomposes the multipart into its original components, the XML control entity and the content entity aimed for the mobile client, and sends a new XML entity in the response back to the push initiator that indicates whether the submission was accepted or not.

From the push initiator's perspective, other operations include cancellation of a previously submitted push message, status query of a previously submitted push message, and the possibility to query the push proxy gateway about a specific client's capabilities (in order to be able to adapt the content to be pushed). Client capabilities may be derived from

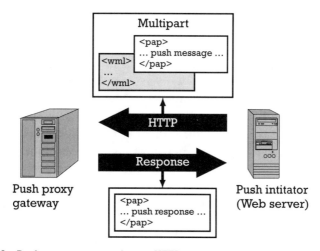

Figure 6.3 Push access protocol over HTTP.

the capabilities negotiated during session establishment [1] or from the user agent profile information [3], if available. The push initiator can also include assumed client capabilities as a third entity in the multipart sent in a push submission, and if they do not align with the actual capabilities of the client, the push proxy gateway cancels the push and notifies the push initiator about it. This procedure relieves the push initiator from the burden of querying the push proxy gateway about the capabilities if it can make a good guess about them and still wants to make sure that they are correct. This can be the case when the push initiator has made a capabilities query earlier, or when it attempts to push content that is supported by a wide range of devices.

If requested by the push initiator, the push proxy gateway can initiate a message to be sent to the push initiator about the final outcome of the push submission (delivered, expired, etc.). If a push initiator wants to use this feature, it must be able to act as a server since it then needs to be able to listen to a HTTP POST request from the push proxy gateway containing the message, and return an XML entity indicating if the message was understood. All other operations in the push access protocol rely on the push proxy gateway's ability to act as a server, and hence, the push initiator acting as a client.

6.4.4 The push proxy gateway

The push proxy gateway [4] is the entity that connects the Internet domain with the wireless network. On the Internet side its responsibility is to serve various requests sent by the push initiator and to notify it about the outcome of push submissions. The push access protocol implicitly specifies most of these responsibilities. The push OTA protocol defines how the push proxy gateway and the mobile client communicate with each other over the wireless network (see Section 6.4.5). Figure 6.4 illustrates the principle of the push proxy gateway architecture.

The protocol stack found to the right under the push proxy layer is a standard Internet stack, including both HTTP client and server. The security (SSL/TLS) layer is optional, but most commercial implementations will likely support these protocols in order to provide necessary measures of security.

The standard WAP protocol stack is found to the left under the OTA layer. Support for WTLS, the security protocol defined by WAP, is optional. But, once again, it is likely that many products will support it. Security is further discussed in Section 6.5.

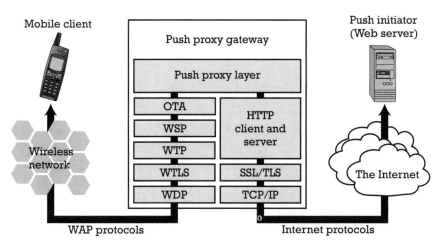

Figure 6.4 Push proxy gateway.

6.4.4.1 The push proxy layer

The push proxy layer found in the upper part of Figure 6.4 enables communication between the Internet and WAP protocol stacks. It contains functions needed to translate between these protocols as well as other functionality. Some of the most important functions are summarized here (some of them are not required by the WAP standard, but are likely to be found in real-life implementations).

Parsing of control information   The first step involved when a push proxy gateway receives a push submission is parsing of the control information provided by the push access protocol, and possibly notifying the push initiator of errors in that information. If free from errors, the push proxy layer uses the control information in the subsequent processing of the push.

Content transformation   The push proxy gateway usually needs to perform some kind of transformation on the content to be pushed to the mobile client. The most common transformations rely on using wireless binary XML (WBXML) to encode, for instance, WML into a compact binary format suited for over-the-air delivery. The push proxy gateway may also perform other "intelligent" transformations based on its knowledge about what content types the mobile client supports. It may, for example, convert a JPEG image into a bitmap image if the client only supports the latter format.

Session management   When using the connection-oriented mode [1] provided by WAP, the push proxy gateway needs to be able to manage the sessions established between it and the mobile clients.

Client capabilities   As mentioned in Section 6.4.3.2, the push proxy gateway needs to be able to find out about a specific client's capabilities. These capabilities could either be derived from the capabilities negotiated during session establishment, or from a server dedicated to providing capability and preference information. The UAProf [3] activity in the WAP Forum addresses the second possibility by allowing the push proxy gateway to access a composite capabilities/preference profiles (CC/PP) server (also known as a CC/PP repository) that holds information about clients' hardware, software, and subscriber preferences.

Store-and-forward   The connection-oriented mode provides a means to use confirmed push; that is, the push proxy gateway will become aware of whether the push successfully reached the addressed user agent in the mobile client or not. Given the push initiator's possibility to stipulate time constraints in a push submission, this implies that a push proxy gateway also needs to implement store-and-forward functionality if retransmissions are required. It also needs to implement retry schemes in the case of a prior unsuccessful delivery, unless the push proxy gateway is notified when the mobile client becomes available again. The push proxy gateway can notify the push initiator about the final outcome of a confirmed push, if requested.

Prioritization   A push initiator may specify three different levels of priority for a push message. It is left to the push proxy gateway how to use this information and what actions to take in order to make use of the levels. For example, a high-priority push may be assigned a low-latency bearer with high bandwidth and precedence in the queue of outgoing messages.

Address resolving   An important feature is the push proxy gateway's ability to resolve the address(es) supplied by the push initiator. The push initiator may provide a network address or a logical address for the mobile client using a well-defined addressing scheme. A network address (for example, "WAPPUSH=+123456789/TYPE=PLMN@ppg.operator.net") can be used by the push proxy gateway directly, but if a logical address is specified, the push proxy gateway must be able to find the corresponding network address for that particular subscriber. For example, if the address

"WAPPUSH=bo.larsson@ausys.se/TYPE=USER@ppg.operator.net" is specified by the push initiator, the push proxy gateway needs to be able to convert the user-defined identifier ("bo.larsson@ausys.se") to an MSISDN if a
bearer like SMS is used, or an IP number if, for example, GPRS or CDPD is
used.

Addressing of multiple recipients   It is possible to specify multiple recipients in a push submission, but a push proxy gateway implementation may
go one step further and support the possibility of using logical addresses for
multiple recipients. For example, to target a push message to all the
employees of an AU system, the user-defined identifier "all@ausys.se"
could be used. Such addresses would thus imply that the address is mapped
to a well-known set of subscribers. It is also possible to map an address to an
unknown set of subscribers; for example, a push addressed to the user-
defined identifier "knightsbridge@london.uk" could notify the subscribers
in the London/Knightsbridge area that there is an ongoing sale at Harrods.
The latter example would require that a bearer supporting cell-broadcast is
available.

6.4.4.2   Management functions

There is also a range of management functions from which the push
proxy gateway operator would benefit. None of these is standardized,
and implementations will surely vary in this respect. Some examples
include:

Push initiator management   The push proxy gateway operator needs to be
able to control the access to the push proxy gateway so no unauthorized
push initiators are provided access (this is also discussed in Section 6.5).
This is in order to avoid unsolicited pushes (i.e., spam). In addition, the
operator will likely want to be able to control what content a certain push
initiator is allowed to push, what delivery time constraints it may place, if it
is allowed to address multiple recipients, if it is allowed to only address a
certain group of recipients, what bearers it may use, which priority levels it
may use, which client-side applications it may address, etc.

Subscriber management   Each subscriber has individual preferences and
usage patterns. For example, individual preferences may include users
having different opinions on what bearer to use since different bearers
imply different costs and performance, and some users may want to limit
the type or size of the pushed content. Individual usage patterns could, for

instance, mean that some users may choose to only have pushes delivered to them during certain hours, for example, only during business hours. In order to allow the operator to better meet users' needs, it must be possible to do certain settings in the push proxy gateway. Since it might be too burdensome for the operator to do these settings for each subscriber, the possibility to specify them for groups of users (for example, based on the type of subscription, or a set of profiles the users can choose between) would likely serve as a fair compromise in many situations.

Network bearer management   Both management entities mentioned above imply that the operator needs to be able to control how the available bearers are to be used. In addition, it also needs to be able to institute policies that control the bearer selection mechanisms when the push initiator does not specify any particular bearer in the push submission. These policies could, for example, be based on costs, availability, reliability, and/or traffic load patterns.

Even though the push proxy gateway and the method proxy gateway are considered to be separate logical entities, there are functions they may share in an actual implementation. Examples include encoders that transform (e.g., WML into its binary form), session management functions (a WSP session can be used for both pull and push), and perhaps the subscriber and network management entities as well.

6.4.5 Push OTA protocol

The push OTA protocol [5] is a thin layer above the WSP that basically extends the facilities provided by WSP with push-specific necessities. The protocol supports both connection-oriented and connectionless modes (i.e., it enables push both within and outside a session). The connectionless mode only supports unconfirmed push, that is, an unreliable "fire-and-forget" delivery method. Connection-oriented mode supports both unconfirmed and confirmed pushes. Note that a confirmation in this context implies that the addressed user agent in the mobile client has accepted the pushed content for further processing. It does not only inform the push proxy gateway whether the mobile client has received the push or not, which a bearer level confirmation would imply.

6.4.5.1 Session initiation application

In WAP, the client always creates the sessions (i.e., the client contacts the push proxy gateway and requests for a session to be created). So, if a push proxy gateway wants to push something to a client in a confirmed manner, which requires a session, and a session does not exist, there would be nothing the proxy could do if this problem were not addressed by the push framework. The OTA protocol provides a solution to this problem by introducing the session initiation application (SIA). The SIA makes it possible for the push proxy gateway to send a request to the client in an unconfirmed manner (without any session) that instructs the client to create a session for a specific user agent and bearer.

6.4.6 Mobile client behavior

SIA is an example of a client-side application that enables server-initiated sessions. A mobile client can have several other push-enabled applications installed, including both user agents defined by WAP (WML and WTA user agents; see Chapters 2 and 4, respectively) as well as other applications (for example, a proprietary mail application or a game).

6.4.6.1 Application dispatching

When something is pushed to a mobile client, the push initiator is not only interested in making sure that it reaches the correct client, but also the correct application in the client. This is taken into consideration by using an *application-ID* that uniquely identifies a particular application in the mobile client. Each application in the client is assigned a URI that is used as an application-ID, but a registered numeric identifier (assigned by the WAP Forum) may also be used for improved over-the-air efficiency.

When the connection-oriented mode of WSP is utilized, the client informs the push proxy gateway which application-IDs it accepts when a session is created. The push proxy gateway can then use that session as a means to push something to those applications, each distinguished by its application-ID. This provides the push proxy gateway with knowledge about installed client-side applications so that bandwidth is not wasted if a push initiator tries to push something to an application that is not installed. There may be several sessions established to make different combinations of applications, bearers, and capabilities possible.

If the connectionless mode is used, two well-known client-side ports are used instead. One of the ports is for secure push; that is, it requires a WTLS connection to be established between the mobile client and the push proxy gateway. The other port is used for nonsecure push. The client is responsible for dispatching each push to the correct application based on the application-ID, or silently discarding the push if the addressed application is not available in the client.

6.4.6.2   Application behavior

Upon reception of a push message, the receiving application is responsible for handling the pushed content. If neither the content type nor the application itself specifies any push-specific semantics for the received content, the content will be placed in the client's cache. If no cache is implemented, the pushed content will be discarded. For example, WML does not specify any push semantics, and the WML user agent does not specify any push behavior for WML. So, if WML is pushed to the WML user agent, it will not be presented to the user but placed in the cache.

Now you probably ask yourself how this makes sense. Should it not be possible to use, for instance, WML in push services? Of course, it should, and as you will soon see, it is.

One of the reasons for choosing the design described above is that it is potentially too user intrusive to start executing or rendering arbitrary content when it is received by the mobile client via push. If the user is not provided with a means to choose when a pushed service should be executed, and in which order if there are many of them, he or she would probably get tired of services popping up in the display at the wrong moments. In addition, neither the push initiator nor the push proxy gateway has any knowledge about the mobile client's state (for example, another service may be executing in the client), so the burden of assuring a good user experience cannot be placed on those entities.

Another restriction that manifests itself is the hardware capabilities of thin clients. Suppose that multiple pieces of arbitrary content is pushed to a mobile client, but not looked at or acted upon by the user, perhaps because another service is running or simply because the mobile client is left unattended. Then the content would need to be queued in the client so that the user would not miss anything. This works fine on a desktop computer with lots of hardware resources, but hardly on a thin mobile client with limited memory.

The bottom line is that the user, to some extent, must be in control of pushed content, and it must also be possible to keep the pushed content small in size so it can be queued in the mobile client if necessary. This is addressed by introducing new content types called service indication and service loading.

6.4.7   Service indication

Just like WML, service indication [6] is an XML-based content type that may be binary encoded for efficient over-the-air transmission. It is used to inform the user of a mobile client that an external asynchronous event has occurred, and it provides the user with the possibility to start a service to handle that event. For example, a voice mail notification may be sent to the mobile client accompanied with the opportunity to start the voice mail service. This is accomplished by allowing a service indication to convey a small text message and a URL to the mobile client. The message is displayed to the user, and he or she is provided with the option to load the service indicated by the URL directly or to postpone the service indication for later handling. The service indication concept is well suited to allow postponed indications to be accessed from an inbox—how this is accomplished, however, is left to the implementation. Figure 6.5 illustrates the principle.

The service indication is delivered from the push initiator residing on a Web server to the mobile client via the push proxy gateway (the WAP gateway contains both a push proxy gateway and a method proxy gateway shown in Figure 6.5). When the user chooses to load the service indicated by the URL, the client pulls it from a Web server via the method proxy gateway. The example in the figure assumes that the push initiator and the service to be loaded both reside on the same Web server, but service indication allows them to be residing on separate Web servers as well.

The "pull line" in Figure 6.5 is dotted because the request for the content may not reach the origin server since the content to be loaded may already be present in the client's cache. Content may end up in the cache if it has been accessed earlier, but that is usually not the case when service indication is used since it most often is used to inform the user about something new. One way of placing content in the client's cache before the service indication is presented to the user is to use multipart

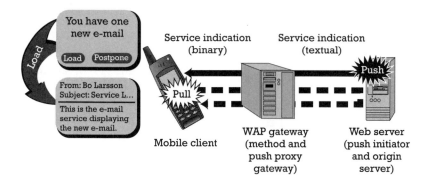

Figure 6.5 Service indication.

(not to be mixed up with the multipart sent between the push initiator and the push proxy gateway). If the content indicated by a service indication, for example, a WML deck, is placed as the first entity in a multipart conveyed to the mobile client and the service indication is placed as the second entity, then the WML deck will be cached before the service indication is presented. So when the user chooses to load the indicated service, the content is readily available in the cache, which will improve the user experience. But remember, given what was said in Section 6.4.6.2, a priori knowledge of the client's capabilities is recommended if this approach should be used. If a thin client is addressed, it is usually better to only send the service indication.

Service indication also provides some additional features to improve its usability. These include deletion and replacement of previously submitted service indications, resolving race conditions, and the possibility to specify the level of user-intrusiveness (controls when the service indication should be presented to the user if the client is busy when it is received). It is also possible to specify when a service indication expires and thereby should be automatically deleted.

6.4.8 Service loading

Service loading [7] is also an XML-based content type, and just like service indication, is used to instruct the mobile client to load content indicated by a URL into a clean user agent context. The difference is that no message can be presented to the user, and the indicated service will be loaded without any user intervention at all. Hence, the user will experience the indicated service as if it were pushed and executed or rendered directly. It is also possible to use the content type to instruct the client to preemptively place the content indicated by the URL in the cache.

This is directly contrary to what was said in Section 6.4.6.2, and the content type should therefore be used very carefully. The content type is first and foremost intended to be used in services that require some kind of user interactivity where the user would find it odd if he or she had to confirm every push.

The push initiator management functions discussed in Section 6.4.4.2 included the possibility of controlling what content types different push initiators should be allowed to include in a push submission. Since service loading is a content type that can be misused, it is thus a splendid example of a content type of which it should be possible to restrict the use.

6.5 Security aspects

Security is an extensive topic, and in-depth knowledge is often required to understand its implications. As this chapter is about push, this section is limited to introducing some basic security aspects to be considered when delivering content using push. A detailed discussion of security issues in WAP can be found in Chapter 7. However, while the focus is on push, much of the reasoning also applies to pull. Since the push framework defines delivery mechanisms to be used both on the Internet and in the wireless domain, security considerations need to be addressed in both cases.

6.5.1 Internet security

A range of security protocols to be used on the Internet is already widely available, allowing push initiators and push proxy gateways to communicate in a safe manner. The secure socket layer (SSL) is the most frequently used security protocol on top of the transport control protocol/Internet protocol (TCP/IP), especially in conjunction with HTTP (often referred to as HTTPS). It provides mechanisms for authenticating both servers and clients, encryption to ensure data confidentiality, and message authentication codes to ensure data integrity. If transport protocols other than HTTP will be accommodated by the push access protocol in the future, protocols like secure/multipurpose Internet mail extensions (S/MIME) or Internet protocol secure (IPsec) may be other qualified candidates.

The features provided by protocols like SSL might in some cases be superfluous. For example, if the WAP gateway is only connected to a corporate intranet, there might not be a need for a security protocol, or the means provided by HTTP itself (for example, HTTP basic authentication, a simple user/password mechanism) might suffice.

6.5.2 WAP security

In WAP, the WTLS [8] protocol provides the same functions as listed for SSL. As a matter of fact, WTLS is derived from the transport layer security (TLS) protocol, which in turn is based on SSL version 3.0. WTLS is optimized with respect to the number and size of the messages sent over the air, and it can also run on top of an unreliable transport protocol.

6.5.3 End-to-end security

So then, are there no hindrances to establishing an adequate security relationship between a push initiator and a mobile client? Well, that depends on the situation, and especially on which type of service is to be implemented. While, for instance, SSL may be used on the Internet and WTLS in the wireless domain, to provide sufficient security in each specific case, the push proxy gateway needs to be able to translate between these protocols and possibly also perform various transformations on the content. In doing so, the security chain between the push initiator and the mobile client is broken.

The lower part of Figure 6.6 attempts to illustrate that end-to-end security can only be accomplished when the mobile client communicates with a WAP server. The WAP Forum has considerably improved end-to-end security in the WAP 1.2 specifications released in November 1999.

An end-to-end solution is most often the only viable one when services like banking and e-commerce are brought about. However, transitive trust (also known as delegated trust or hop-by-hop security) is an acceptable solution for most other services.

6.5.4 Transitive trust

Transitive trust can be established if the push proxy gateway, or rather the push proxy gateway operator, can be considered trusted by the user

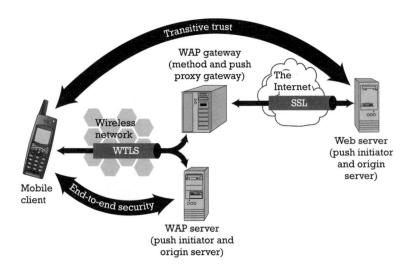

Figure 6.6 Security.

of the mobile client. Among the features provided by the security protocols, authentication is one of the most important features in this respect. It makes it possible to verify that a message actually originates from the source from which it claims to originate. SSL makes it possible to authenticate push initiators in the push proxy gateway (using X.509 certificates), enabling the push proxy gateway to maintain a rigid access control. WTLS provides a means to authenticate push proxy gateways and WAP servers in the client, and vice versa. So, if the user knows that the push proxy gateway only accepts pushes from push initiators whom he or she trusts and the push proxy gateway can be authenticated by the client, then the user knows that the content being pushed originates from a trusted push initiator.

In order to accommodate transitive trust, the push framework introduces a couple of push-unique features (i.e., features that are not available for pull). These provide a push proxy gateway with a means to indicate to the mobile client that the push initiator has been authenticated and if the content can be trusted.

6.6 Making it happen

The concept of push in the mobile environment is not totally new, but the means available until today have certainly acted as an impediment to the inventiveness among operators and third-party service providers. This becomes fairly obvious when one compares the push services offered and their ability to grasp business opportunities in other areas. WAP has scored an unparalleled success in the wireless data community, and with push entering the scene, we will likely see a plethora of new services evolve. As always, when a new technology is introduced or made more powerful, some of the services will score tremendous success, while others will fall flat as pancakes. It is after all, at least to some extent, a new territory.

Finding the motive power and avoiding the pitfalls when push is introduced is by far not an easy task, but a challenging and interesting one, at least in my humble opinion. Unfortunately, it would require much more than a chapter to provide a good analysis; an entire book is needed. So let us only look very briefly into this in order to raise some concerns before some examples of push services are given.

6.6.1 Understanding customer value

A key driver in launching successful services is without a doubt customer value. In creating customer value, one should always be guided by fundamentals like convenience, efficiency, flexibility, simple to use, etc. When pull-based services are created using WAP, we can learn a great deal about the intrinsic value of these fundamentals from the Internet community. With push it is somewhat different since push technology is not as widely deployed as pull technology on the Internet as of today. It will probably take some time before we understand the fundamentals for push just as well as we understand them for pull. The situation is further complicated by the fact that we now have two technologies that shall collaborate, that is, push and pull. Thus, one should not consider the fundamentals for push and pull separately. Rather, it is important to be able to see how they interact with each other in order to be able to launch successful service concepts.

6.6.2 Understanding the value chain

When understanding the mechanisms for creating customer value, the next step is to find out how to make money out of it, both with respect to attracting new customers and retaining existing ones. It is not only the number of customers that should be considered, but also their tendency towards using the services offered.

An important decision for the operator is how it should position itself in the value chain. Should it act as a full-fledged service provider, only as a pipe providing network capacity, or somewhere in between? While the following reasoning is applicable to both push and pull, it is important to remember that the push proxy gateway operator and a push initiator likely need to establish some sort of business relationship in order to provide the push initiator access to the push proxy gateway. So, when push is brought about, the operator is provided with larger flexibility when positioning itself in the value chain since it can more effectively control what services third-party service providers should be allowed to deliver.

Without WAP, the operator that runs a mobile network traditionally controls almost the entire value chain for mobile services. Third-party alliances are not very common, even if they have become more frequent during the last couple of years. This scenario will most probably change rather dramatically for both parties mentioned when WAP enters the scene. Using the Internet as a service platform opens new possibilities for third-party service providers to take part in the value chain at different stages. Third-party service providers will be able to create WAP services, put them on the Internet, and thereby make them available to millions of subscribers. They will even be able to create complete suites of services and thus also affect the operator's role in bundling services.

With the magnitude of new services that WAP will make available, users will become increasingly aware of the utility they provide, and network operators are unlikely to be able to serve all of their customers with self-made services that attract each and every one of them. Their position in the value chain should make it possible for them to differentiate themselves from their competitors and have flexibility enough to respond to new preferences among their customers and changes on the market for mobile services in general.

6.6.3 Making the money

No matter to what degree the operator decides to cooperate with third-party providers, it will still enjoy an increased network utilization, which will have a positive impact on the earnings. Third-party cooperation ought to be considered in order to maximize that utilization and to provide a well-adapted mix of services that allows the operator to differentiate itself from its competitors and attract new or underdeveloped market segments as well as retaining existing ones. This will reduce churn and improve customer loyalty, and thereby pave the way for increased revenues.

Independent of what business model the operator uses for pull, it may need to adopt other models for push. For example, when the user pulls content from a server, it might be feasible to charge for the bearer utilization since the user has a priori knowledge of the transaction. That model might not be very good for push if the user cannot control the number of messages sent, and thus not be able to control the costs incurred. A possible solution to the problem could be a flat-rate subscription, where the user either pays for push capability in general or a fixed amount for each separate service to which he or she subscribes.

With push it is also possible to use a reverse billing scheme. A service provider (push initiator) may pay a fee (fixed or variable) to the operator for accessing the push proxy gateway and using the bearer network. When a user subscribes to a push service, he or she pays a subscription fee to the service provider instead of to the operator. The operator might, however, bill the subscription fee for the user's convenience, but that is another issue. One way for the user to avoid the subscription fee would be to allow advertising, for which in turn the service provider can charge the advertiser.

6.6.4 Some examples of push services

Here are some examples of push services; the list could be much longer. The first two paragraphs provide examples of services that could be implemented using WAP 1.2, while the last two paragraphs try to illustrate what the future might have to offer.

The first step towards push in WAP is to outline a migration path for existing services, for example, SMS-based services. A faithful old servant is notifications, primarily voice mail notifications. Such services can easily be converted, and enhanced to WAP-using service indication. Other legacy SMS services subject to migration include traditional information services like news, sport results, stock quotes, weather, etc., and also more ingenious ones like jokes (which should not be underestimated—jokes over SMS have become one of the more popular services in Norway, for instance).

The next step is to integrate push applications with existing systems. A typical example is integration with a corporate exchange server, allowing contacts, e-mails, and meeting requests to be pushed to the mobile client. Another example is integration with an application that monitors an automated assembly line. Using a wireless device capable of receiving pushes, the technician on duty could be notified about errors wherever he or she is.

There are several examples relating to banking and e-commerce. For example, order and pay a flight ticket and have it pushed to your mobile device in the form of a virtual ticket. When you arrive at the airport, you simply enter the flight operator's Bluetooth zone where a virtual boarding card is pushed to your device, you put the luggage on the conveyor belt, and you are ready for boarding. It could also be possible to periodically push a transfer of e-money to the device to be stored on a smart card, readily available for paying for the flowers that you ordered for your significant other from the florist's WAP home page.

There is an ongoing activity in the WAP Forum relating to telematics that, among other things, include positioning. If the position of the mobile device is known, it would be possible to create push services that, for instance, inform you about sights in the different areas you visit on your vacation, and, if you travel by car, you could also be provided with driving directions in order to not miss the scenic routes. Another example is a taxi company that uses the position information to manage its fleet by pushing driving orders to its drivers.

References

[1] Wireless Session Protocol Specification, Version 5—November 1999, WAP Forum, www.wapforum.org.

[2] Push Access Protocol Specification, Version 8—November 1999, WAP Forum, www.wapforum.org.

[3] User Agent Profile Specification, Version 10—November 1999, WAP Forum, www.wapforum.org.

[4] Push Proxy Gateway Service Specification, Version 16—August 1999, WAP Forum, www.wapforum.org.

[5] Push OTA Protocol Specification, Version 8—November 1999, WAP Forum, www.wapforum.org.

[6] Service Indication Specification, Version 8—November 1999, WAP Forum, www.wapforum.org.

[7] Service Loading Specification, Version 8—November 1999, WAP Forum, www.wapforum.org.

[8] Wireless Transport Layer Security Specification, Version 5—November 1999, WAP Forum, www.wapforum.org.

CHAPTER

7

Contents

Wireless Application Protocol Security

Simon Blake-Wilson, Robert Gallant, Hugh MacDonald, Prakash Panjwani, and Greg Sigel

7.1 Introduction

Technological advances have brought commerce into the home, extended communication beyond the wired confines of the home, and enhanced the capabilities of wireless devices far beyond those limited to pocket electronic organizers and cellular voice. People and businesses have become accustomed to the availability of quick and easy communications and are performing all sorts of tasks using their wireless devices.

The introduction of wireless data initiated the convergence of telecommunications, the Internet, and electronic commerce. WAP companies have joined forces to expand the limits of wireless e-commerce, while adhering to the demands of the various end-user communities. The security aspects of WAP permit people and businesses to conduct

143

their confidential and sensitive transactions wirelessly with confidence that the data will remain unaltered during transmission and that only the intended recipients will have access to that data.

These cases illustrate the kind of transactions that need security.

7.1.1   Case 1

The president of a public company, rushing out the door to a board meeting to present the quarterly report, realizes that she does not have the quarterly figures readily available. Passing by the Chief Financial Officer's (CFO) office, she asks him to e-mail the figures to her mobile account so she can read the information on her PDA while in transit to the meeting. Needless to say, neither the CFO nor the president wants anyone except for the president to be able to read the message, nor can they risk having any of the information mutated.

7.1.2   Case 2

An active day trader on the stock market needs to keep track of the value of his stocks regardless of where he is, so he uses his two-way pager to grab stock quotes from the Internet when he is on the road. Whenever the price plummets, he immediately purchases, and similarly when the prices of his shares soar, he sells them off to cash in on his good fortune. This trader doesn't want his stock portfolio to be available to the public. He needs to keep the selection of stocks that he monitors private. He also needs to know that the stock quotes that he receives and responds to do in fact come from a trusted source and that the values received by both parties (from and to the broker) exactly match those that were sent.

7.1.3   Case 3

After being informed of a golden opportunity to close a sale in Vancouver by the end of business today, a saleswoman in Toronto leaves the office in a rush for the airport. In the taxi she accesses her favorite travel site via the WAP browser on her mobile phone, checks the flight availability, and reserves a seat on the 10:31 A.M. flight using her credit card. In reserving this ticket while riding to the airport, she wants her credit card information to remain secure, and she wants her confirmation number to be accurate. She also wants to be sure that she is communicating with a valid and trusted ticket-selling agency.

7.1.4 Case 4

Coming home on the bus, after purchasing a new car, a man realizes he has just written a check that will almost empty his checking account and that his monthly rent check will be cashed first thing the next morning. Within seconds, he connects to his bank using his palmtop and checks the balance in his checking and savings accounts. Noticing that he has enough money in his savings account to cover the expense, he transfers the difference to his checking account. In order to initiate the communication, it is essential to verify that the two parties communicating are the owners of the account information and the bank. During this transaction, the man in question needs to know that his account information will be kept private and that the cash value that his palmtop receives is the same value that the bank sent. That is to say, there has been no change to the information on its way from the bank to his palmtop. In summary, when he sends the request to transfer money from one account, he needs to know that only he can access and manipulate his account, that the request remains unchanged from when he sends it to when the bank receives it, and that only the bank can interpret the request that he has sent. The bank also wants to be sure of whom it is dealing with, the amount to be transferred, and that the man transferring the money cannot later deny having done the transfer.

In each of these examples, the need to keep some of the information private and to authenticate both the entities in the communication as well as the data transferred is clear. WAP is providing the wireless community with the opportunity to securely provide applications, including electronic commerce, stock trading, two-way pager messages, and banking. These applications require security to ensure their proper use and to protect the end user from a malicious attack and the provider of the device/service from liability.

This chapter describes cryptographic functionality built into WAP via the WTLS specification to provide the security and authentication required to perform these and many other wireless communications with confidence.

The remainder of this chapter is organized as follows. Section 7.2 provides an overview of cryptography. Section 7.3 describes the challenges faced when implementing cryptography in a wireless environment. Section 7.4 discusses the WTLS specification. Section 7.5 contains conclusions, and finally a bibliography is given for those who want to know more. A list of the acronyms found throughout this chapter is available in the back of the book.

7.2 Overview of cryptography

With the increased use of electronic media for storage of information, there is an increased need to provide electronic security. Any information that is to be communicated from one entity to another is exposed to the possibility of an attack by an adversary. With a hard copy of information, a trusted courier and a hand signature are sufficient for reasonable security. When the item to be communicated is stored electronically, it is easy to copy, alter, read, and insert fraudulent data, or intercept data if they are sent without some form of further protection. Cryptography has taken the role of securing electronic data and the entities involved in electronic communication.

There are several services required of cryptography in order to ensure that the communication is in fact secure.

1. *Data confidentiality*. Quite often the first aspect that springs to mind when cryptography is discussed, this is the act of keeping secret the data that are to be communicated, so that only people with the appropriate access may see the data. The need for data confidentiality is seen in the first example, where the CFO and the president of a company were detailing sensitive information in an e-mail message to be sent over the Internet.

2. *Data integrity*. This refers to the task of ensuring that data exchanged during communication remain unaltered. An example of data integrity is the need of the stock market player to be sure that the values of the stocks that he receives are in fact the true value of the stock as sent by his trusted source.

3. *Data origin authentication*. Often underestimated, this may be the most important cryptographic service in many applications. It may be important to be able to verify the source of the data received during communication, to avoid the possibility of someone inserting an invalid response in the middle of an established connection, or to avoid communicating with adversaries who have falsely identified themselves. In the stock market example, the need for data origin authentication is shown, as the trader needs to verify that the values he receives are from a source which he trusts.

4. *Device or entity authentication*. One or both of the entities involved may need to verify that the other entity with whom they are

communicating are who they claim to be. In the example of the saleswoman purchasing an airplane ticket on her way to the airport, the need for entity authentication can be seen. Since she is buying an airplane ticket, she needs to know that the entity on the other end of the transaction is in fact a valid seller of airplane tickets. An example involving device authentication would be a two-way pager system. When a message is sent from the pager, the paging network receiving the request needs to be able to identify the pager as having a valid contract.

5. *Nonrepudiation.* After an exchange has occurred involving an agreement or a transmission of data, it is often important that the parties involved not be able to deny having entered into the agreement, or having sent the data in question. The airplane ticket seller is interested in the nonrepudiation feature of cryptography when the saleswoman agrees to purchase the ticket. The seller needs to be sure that sometime in the future the saleswoman cannot deny having purchased that ticket and demand a refund.

The example of the man coming home from purchasing a car exhibits each of the features of cryptography, some of them several times. The man needs to verify with whom he is communicating, so that he is sure he is sending his confidential account information to the correct bank, and thereby needs entity authentication. He also needs data confidentiality when he is sending his account information to the bank so that he may query his accounts. He wants his balance to remain private as well. Both he and the bank need to be sure that the account information that he sends remains unaltered, using the data integrity aspect of cryptography, so that the accounts queried are the correct accounts and that the amounts returned are valid. As a further example of this, the man needs to know that the account balances he receives are the values that were sent by the bank. When he is receiving the balance of his accounts, he needs to be sure that the bank was in fact the sender and that nobody inserted a value during the communication; in other words, he needs to authenticate the origin of the data. After transferring the required funds from the man's savings account to his checking account, the bank needs to be certain that he will not be able to contest or repudiate this transfer in the future. Although this is not an exhaustive list of the features needed for this example, it shows that some or all of the five basic cryptographic services may be needed in combination.

In order to achieve the five basic services, two different types of cryptography are commonly used. These types are symmetric-key and public-key cryptography. Both methods have strengths and are particularly good in accomplishing certain tasks, but they both also have drawbacks and therefore are not perfect in all applications. Thus, the two are used in combination to gain the maximum benefit with the least cost.

7.2.1   Symmetric-key cryptography

A symmetric-key system is one where there is a single secret key known to both the entities involved in the communication. This is known also as a secret-key system. The main advantage of symmetric-key systems is they tend to have very fast implementations and are therefore good for securing large packets of data. The drawback, however, is that each pair of entities must share a secret key known only to them, as the same key is used to both encrypt and decrypt the data. For example, if a system had 100 people in it, each of whom might need to communicate confidentially with each of the others, a symmetric-key system would require close to 5,000 keys, and all of these must be kept secret. Establishing and maintaining the secrecy of these keys are hard.

Well-known symmetric-key systems include the United States government's Data Encryption Standard (DES) and its successor, the Advanced Encryption Standard (AES), which is due to be selected in 2000.

7.2.2   Public-key cryptography

In contrast, a public-key system is one where each user has two keys, one known as a private key which is kept secret, and the other known as a public key which is public and made available to everyone in the system. To encrypt a message to send to Alice using a public-key system, Bob looks up Alice's public key, encrypts the data using this public key, and sends the resulting ciphertext to Alice. Alice uses her private key to decrypt and retrieve the data. An important feature of a public-key system is that as a result only 200 keys are required in a system of 100 users, and only 100 need to be kept secret. A clear advantage of a public-key system is that there are far fewer keys to manage. An additional feature of public-key systems is that, unlike secret-key systems, they are able to provide nonrepudiation—if Alice signs a message using her private key, Bob can check that the message came from Alice, and he can prove this legally since only Alice knows her private key. The main drawback of public-key systems is that implementations tend to be computationally intensive compared to secret-key systems.

The relationship between a private key and a public key in a public-key system seems at first like magic. Public-key systems are made possible by the mathematical idea of a trapdoor one-way function. A one-way function is something that takes a value as input, processes that value, and outputs the result. When given only the result of this operation, it is extremely difficult to determine what the original input value was. Thus, anything encrypted using a true one-way function would be secure from prying eyes, as it cannot be reversed. Unfortunately it would also be kept secure from the intended recipient. This is where trapdoor one-way functions come in. These functions take a value as input and process it in much the same way an ordinary one-way function would, and output the resulting value. The difference in this case, however, is that with certain knowledge the original value can be restored (i.e., there is a trapdoor through which we can retrieve the original value). To put this in cryptographic terms, to encrypt a message using a trapdoor one-way function, the sender would process the message using the function and send the output value to the recipient. The recipient, knowing the trapdoor, would process the encrypted text to restore the message to its unencrypted original form.

The security of public-key systems is based on trapdoor one-way functions. The majority of current public-key schemes are based on trapdoor one-way functions that have one of three hard mathematical problems forming the underlying security. By name, these problems are the integer factorization problem (IFP), the discrete logarithm problem (DLP), and the elliptic curve discrete logarithm problem (ECDLP).

7.2.2.1 Integer factorization problem

The RSA scheme, probably the best known of all public-key systems, is based on the integer factorization problem. Given an integer n which is the product of two large prime numbers, the integer factorization problem is to factor n to recover these primes. The integer factorization problem is believed to be hard. This means multiplication is a one-way function—given two primes it is easy to multiply them together to get their product n, but it is hard to reverse the process and recover the prime factors from n. The largest reported value factored presently is a 155-decimal digit number (a 512-bit RSA modulus)—the project used almost 300 computers and took about 7 months. With a concerted effort, distributed over the Internet, it is estimated that the amount of time to break a

512-bit RSA modulus could be reduced to a week. As a result, 1024-bit numbers are today considered to provide a reasonably secure basis for the RSA scheme.

7.2.2.2   Discrete logarithm problem

Given a large prime p, an integer g between 1 and $p-1$ known as the base, and another integer y between 1 and $p-1$, the discrete logarithm problem is to determine a value x for which the following is true:

$$g^x = y(\bmod\ p)$$

This problem is known as the discrete logarithm problem because it is analogous to the logarithms we meet in high school. To illustrate the problem, look at the small example with $p = 7$, $g = 3$, and $y = 4$. In this case, we can compute the answer $x = 4$ by trial and error:

$$3^1 = 3(\bmod\ 7),\ \ 3^2 = 2(\bmod\ 7),\ \ 3^3 = 6(\bmod\ 7),\ \ 3^4 = 4(\bmod\ 7)$$

Like the integer factorization problem, the discrete logarithm problem is believed to be hard. This means exponentiation modulo p is a one-way function—given p, g, and x, it is easy to exponentiate g to the power x modulo p, but it is hard to recover x from the result. With 1024-bit primes p, they are today considered to provide a reasonably secure basis for schemes based on the discrete logarithm problem. These schemes include the well-known Diffie-Hellman protocol, and the United States government's digital signature algorithm, or DSA.

7.2.2.3   Elliptic curve discrete logarithm problem

The elliptic curve discrete logarithm problem is similar to the DLP, but instead of working in the integers reduced modulo p, the set of points on an elliptic curve are used. In the elliptic curve setting, multiplication of integers becomes addition of elliptic curve points—so the elliptic curve discrete logarithm problem is: given an elliptic curve E (analogous to the prime p), a base point G on E (analogous to the integer g), and another point Q (analogous to the integer y), find an integer x such that:

$$xG = Q \text{ on } E$$

Again the elliptic curve discrete logarithm is believed to be hard. Like in the DLP case, a one-way function is obtained from the elliptic curve discrete logarithm problem—given *E*, *G*, and *x*, it is easy to multiply *G* by *x*, but hard to recover *x* from the result. In fact, the elliptic curve discrete logarithm problem appears considerably harder than either the IFP or the DLP. The hardest problem of this type that has been solved (as of April 2000) is the calculation of a logarithm on a curve where the base point *G* has a 108-bit prime as its order. This calculation required about 50 times more work than the 512-bit RSA modulus factoring effort. To achieve adequate security with schemes based on the elliptic curve discrete logarithm problem, one only needs to work with 160-bit numbers rather than 1024-bit numbers. This means schemes based on the elliptic curve discrete logarithm problem are an order of magnitude more efficient than chemes based on the IFP or DLP—and furthermore this efficiency advantage will grow as computing power increases. As a result, many researchers expect schemes based on the elliptic curve discrete logarithm problem to become predominant. Schemes based on the elliptic curve discrete logarithm problem include the elliptic curve Diffie-Hellman protocol and the elliptic curve DSA, or ECDSA.

7.2.3 Hybrid solutions

The different features of public-key systems and symmetric-key systems—namely, the improved key management and nonrepudiation feature of public-key systems, and the computational efficiency of symmetric-key systems—mean that practical solutions usually employ a hybrid of public-key and symmetric-key cryptography. For example, it is common to set up communications using a public-key system to exchange a shared secret key—this is known as performing key exchange—and then secure subsequent messages with a symmetric-key system using this shared secret key. This solution exploits the features mentioned earlier—use of a public-key system for key exchange deals with key management since parties do not have to share a secret key in advance in order to communicate, and use of a symmetric-key solution to secure messages realizes a low computational overhead. A hybrid solution of this type is employed within WAP by the WTLS protocol.

7.2.4 Cryptographic schemes

Cryptographic systems come in a variety of types depending on which of the goals of data confidentiality, data integrity, data origin authentication, entity authentication, and nonrepudiation they are designed to achieve. Two broad classes are *encryption schemes* and *authentication schemes*.

Encryption schemes are designed to provide data confidentiality. An encryption scheme takes the data in their original form, where they are known as *plaintext*, and performs a transformation upon them to make them illegible to the average person. When the data have been transformed, the output is known as the *ciphertext*. In this way, confidential material may be transmitted across an insecure channel without fear of it being read by an unauthorized party. Upon receipt by an authorized recipient, the encryption scheme can be used to decrypt the ciphertext to recover the plaintext. Encryption schemes can be realized by either symmetric-key systems or public-key systems.

Authentication schemes are designed to provide data integrity and data origin authentication. An authentication scheme takes the plaintext as input and produces an authenticator or signature on the plaintext. The plaintext and signature can now be transmitted across an insecure channel. Upon receipt, the plaintext and signature can be checked to ensure the origin and integrity of the plaintext. Authentication schemes can also be realized by either symmetric-key systems—in which case they are known as message authentication codes—or public-key systems—in which case they are known as signature schemes. Signature schemes have the advantage that in addition to providing data integrity and data origin authentication, they also provide nonrepudiation.

An important cryptographic tool used as a component to build many cryptographic systems—in particular, signature schemes—is a *cryptographic hash function*. A cryptographic hash function takes a long input such as plaintext and outputs a considerably shorter value, which can be thought of as a fingerprint of the input. The hash function is designed so that it is hard to recover input from its fingerprint, and so it is hard to find two different inputs with the same fingerprint. The best-known example of a hash function is the United States government's secure hash algorithm, or SHA-1.

7.2.5   Public-key infrastructures

The astute reader will have noticed a gap in the discussion of public-key systems in Section 7.2.3. Although public-key systems go a long way to solving the key management problem—since they obviate the need for communicating parties to share a secret key in advance—it is still necessary to have a mechanism to make sure public keys are distributed authentically. Otherwise, attackers can substitute a public key of their choice for Alice's public key and therefore decrypt messages intended for

Alice. Mechanisms that distribute public keys authentically are known as public-key infrastructures. The most common public-key infrastructures are based on *digital certificates.*

A digital certificate consists of a message containing a user's identity and their public key, along with the signature of a trusted party known as a certification authority (CA) on this message. Now, when Bob wants Alice's public key, he retrieves her certificate and verifies the CA's signature on the certificate. Once the signature has been verified, Bob knows he has an authentic copy of Alice's public key and he can therefore communicate securely with Alice.

Digital certificates can contain other useful information as well as a user's identity and public key. They can contain, for example, an indication of how the key is meant to be used, an expiration date for the certificate, and a serial number, as well as authorization information such as the user's credit limit or access rights.

When certificates are deployed in practice, there is often more than one CA in the system. In this case users of one CA may need to check the certificates issued by another CA. To solve this problem, the CAs may cross-certify each other—meaning that they issue each other with certificates—or there may be a top-level CA that certifies the keys of low-level CAs. Now to obtain Alice's public key, Bob retrieves a certificate path consisting of a certificate issued by his CA to Alice's CA and a certificate issued by Alice's CA to Alice. By checking both certificates, Bob can obtain an authentic copy of Alice's public key.

Another issue when using certificates is revocation. There will often be times when a user or a CA wishes to revoke a certificate—perhaps because the user's key has been compromised or perhaps because the user has left the CA's organization. To solve the revocation problem, CAs usually issue certificate revocation lists, or CRLs. This is a signed list of the serial numbers of revoked certificates. When Bob wants to obtain Alice's public key, he must in addition check the latest CRL issued by Alice's CA to check that her certificate has not been revoked.

Public-key infrastructures based on certificates are an effective way to distribute public keys. They have been widely deployed in existing public-key systems. The best-known specification for public-key infrastructures based on certificates is the ITU's X.509 standard. This also forms the basis for the IETF's PKI, known as the public-key infrastructure for the Internet, or PKIX.

7.2.6 Summary

Cryptography is an effective way to secure electronic data. Cryptographic systems come in various flavors capable of providing data confidentiality, data integrity, data origin authentication, entity authentication, and non-repudiation. However, in order to provide security against increasingly sophisticated attackers, it is important that solutions using cryptography are designed and implemented with care. Cryptographic solutions designed and implemented without substantial cryptographic expertise are often subject of serious and costly security flaws.

7.3   Security issues in a wireless environment

All cryptographic systems face challenges that must be addressed. Meeting these challenges involves steps like: performing a careful risk analysis to ensure that threats such as disclosure, modification, and replay of messages, as well as denial-of-service have been identified; performing careful specification and analysis of the system to ensure that the threats have been countered; using tried-and-tested algorithms to avoid unforeseen attacks; putting a disaster recovery mechanism in place to deal with compromises; and performing a regular review of the system to ensure its ongoing security.

Wireless devices face some additional challenges in order to provide an efficient and secure solution.

On the security side, the ease of launching attacks against wireless networks must be considered. In wireless networks it is comparatively easy to monitor communications with a minimal chance of being detected. If a credit card transaction, for example, were to take place on a wireless network without encryption, an adversary would only have to intercept the communication in order to have all the information required to go on a shopping spree. In addition, the cellular industry has already suffered from cloning, resulting in vast amounts of fraudulent traffic, which persisted until security was added during the introduction of digital systems. Similarly, if data are sent over a wireless network without authentication, then an attacker may be able to alter, insert, or replace the data sent in a way that serves the attacker's purpose—for example, changing an amount ordered, or the location to which the product should be delivered. The ease with which these attacks can be mounted in a wireless network place additional emphasis on the need to deploy security.

On the efficiency side, the amount of storage, computational overhead, and bandwidth is a concern. While the demand for mobile phones, palmtops, and pagers to become physically smaller is increasing, the need for longer battery life is increasing, as is the need to interoperate with land-based devices of increasing computing power.

Because of their small and decreasing size, wireless devices are often limited in their storage capacity. This restriction is important for several reasons. The small storage capacity requires that the space occupied by the underlying code running the cryptographic services be small. Also, with limited space, large cryptographic keys are undesirable, and the keys used need to be able to be stored efficiently. The best example of a public-key system that uses small keys with efficient storage techniques can be seen in the case of elliptic curve cryptography. The security strength of a 160-bit elliptic curve is about the same as 1024-bit RSA.

Additionally, as a result of the typical size of these devices, the computing power of the processors is severely limited. In order for the cryptographic services to be effective, they must be fast. This reinforces the need for the code to be small, but adds the requirement that the code be extremely efficient. It is desirable to keep the number of computations to a minimum as well. Again elliptic curve cryptography in conjunction with symmetric-key systems lends itself well to the challenge. Because of the small size of the key, elliptic curve cryptographic operations are computed very quickly and require less processor power.

Further, while devices are becoming better at transmitting data, the bandwidth for messages is neither unlimited, nor free. It is preferable to keep the messages as short as possible. Hence, it is desirable to keep the message expansion resulting from any cryptographic procedure as small as possible. With a smaller key, digital signatures for small messages are quite small, requiring less data to be transmitted from the device, which once again implies that elliptic curves are well suited to the challenge of providing security in a wireless environment.

Finally, to meet the efficiency challenges, some degree of convergence is desirable between security services provided to applications by WAP and the security services required for basic cellular communications. It is desirable, for example, to use the same core cryptographic algorithms in both WAP and cellular in order to minimize the size of cryptographic libraries.

7.4   Security in WAP

7.4.1   Introduction

Security in WAP is largely defined by the wireless transport layer security protocol, or the WTLS protocol. In terms of the WAP protocol stack, WTLS defines a layer above the transport protocol layer. It is up to each WAP application to enable the security features available, as they are not enabled by default in a WAP connection. In both spirit and architecture WTLS is similar to the TLS protocol, which is the Internet Engineering Task Force's (IETF) standard for securing Internet browsing, and which is the successor of the de facto Internet security protocol, the secure socket layer (SSL) protocol.

Of the goals of cryptography, as discussed in Section 7.2, WTLS can provide confidentiality, data integrity, data origin authentication, and entity authentication between two communicating applications. WTLS does not, however, provide nonrepudiation services. Nonetheless, WTLS provides a robust security environment, and goes a long way towards solving the security problems of WAP applications and users.

An important distinction when using WTLS is whether end-to-end security or security through a proxy is provided. This will depend on the system architecture in place.

Most initial WAP deployments will have a client↔proxy↔server (content provider) architecture. An advantage of this architecture is that proxy servers can be written to communicate with existing Internet (HTTP/HTML) Web servers. Such proxies allow WAP phones to connect to a simplified representation of familiar Web sites. An important implication of this architecture is that there cannot be a secure connection end-to-end between a client application and a server application. This follows since the WAP security protocols, for example, WTLS, differ from Internet security protocols such as TLS. Thus, the proxy must "unwrap" WTLS-secured data from the client, then rewrap the data using TLS before passing it on to the server. Both client and server must place considerable trust in the proxy, which is typically located within the service provider network. It is possible for client and server applications to negotiate a secure connection at the application layer, independent of any underlying security provided by WTLS. The design of cryptographic protocols, however, is very tricky and subject to subtle yet fatal errors. These possible flaws are a main reason for using the WTLS protocol for security in the first place.

In some cases, application servers contain a full WAP protocol stack, and so client data can be routed directly to the server and vice versa which avoids the need to encrypt and decrypt within the carrier network. Thus, WTLS-enabled servers enable end-to-end security between the client and the application servers. In such cases, the proxy may process unsecured data as usual, but secured packets are passed through the proxy unaltered, as they should be.

The WTLS protocol provides security using a variant of the hybrid-type cryptosystem discussed earlier. That is, a public-key system is used to possibly authenticate the two communicating parties, and to establish a shared key between them. This shared key is subsequently used in symmetric-key systems which provide confidentiality and authentication services on the bulk of data packets transmitted between the parties. Details of this process are discussed in the next section.

7.4.2 The WTLS protocol in detail

The WTLS protocol has a client-server architecture, as illustrated in Figure 7.1. Connections secured by WTLS are always initiated by the client, which should be viewed as the mobile device. Conceptually it is helpful to think of the WTLS component of the WAP stack as a state machine, consisting of a number of components, which are outlined here.

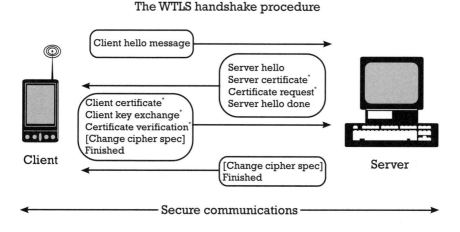

The WTLS handshake procedure

*Indicates optional or situation-dependent messages that are not always sent

Figure 7.1 Client/server architecture of the wireless transaction layer security.

The main component, known as the record layer, contains the state of the current symmetric encryption and authentication mechanisms. Data packets are injected into the record layer, at which point the current cryptographic mechanisms are applied, resulting in a secured data packet, which can then be safely sent on the wireless network. When the other entity receives this secured packet, its record layer processe sit using the same state and mechanisms as the record layer of the sender.

The receiving record layer can authenticate and decrypt the secured data packet, and pass the resulting plaintext data up the WAP communication stack to the relevant application. A pair of entities, communicating for the first time, do not share any secret data, and so the common state of their record layers is the null state. In the null state, the record layer adds some administrative data to a packet, but essentially passes data through unmodified. Four other components use the record layer. By far, the most interesting of these is the handshake protocol component. The handshake protocol is responsible for negotiating a new shared common state for a pair of communicating record layers. It is again stressed that the communicating entities need not share any initial secret data, but the derived key will nonetheless be known only to these two entities. To illustrate a typical handshake, assume that a pair of entities are initiating a connection for the first time. The description here is based on a typical set of messages exchanged between the respective entities during the handshake protocols. Although the initial data are processed by the respective record layers, the initial state is null so the data are essentially passed through unchanged. This exchange of data in the clear is secure as a consequence of using public-key cryptography. The client begins by constructing a message known as the "client hello." Besides serving as a request for a secure connection, the client hello also contains fields for negotiating the public-key algorithm used for establishing the shared key, and for negotiating the symmetric algorithms to be used to protect the packet data. The client hello also contains random data generated by the client to prevent replay and known key attacks.

From the choices listed in the client hello, the server decides which key exchange suite to use, as well as which symmetric suites to use. The choices selected are contained in the "server hello," which also contains some administrative data for the session, as well as random data generated by the server. Since the key-exchange algorithms have now been decided upon, the server may now send its certificate and/or its key-exchange contribution. This information is sent immediately after the server hello. If client authentication is desired, the server may request the client certificate at this time.

Upon receiving the data from the server, the client learns the selected cryptographic algorithms. In particular, the client can respond with its certificate or other public keys relevant to the key exchange mode. At this point the client knows the master secret, which is essentially the key derived from the public-key exchange. This key can be used in the now known symmetric algorithms, in particular, the symmetric authentication algorithm. This is used to generate some client authentication data, which is also now sent. From this point on, all data sent from the client will be encrypted and authenticated using the negotiated symmetric algorithms and shared keys. The server receives the client's contribution to the key exchange, as well as the authentication data. Thus, the server may also calculate the shared master secret and use it in combination with the received authentication data to authenticate the client. The server then generates some data to authenticate itself to the client, and sends this. From this point on, all data sent from the server to the client will be encrypted and authenticated using the negotiated algorithms. Finally, the client receives the authentication data from the server and verifies them, thus authenticating the server. The client and server are now free to exchange application data packets, each of which are protected using the negotiated algorithms. This protection persists for the length of the communication session.

In subsequent communications, the client and server can use the data from previous sessions to accelerate the negotiation of a new secure connection. This process is known as session resumption, and is quite useful in that it typically requires less data exchange and computations and is therefore very fast. There are three other components that use the record layer. Error and warning notification during communications is managed by the alert protocol. The application data protocol manages the process of passing application data through a secured connection. Finally, the change cipher spec protocol manages the messages exchanged when switching from nonsecure communication to secure communication.

7.4.3 Security attributes of wireless transport layer security

Security is a complex area. Many attacks can be attempted on a given cryptosystem, and it is highly likely that at least some of them, though possibly subtle or benign, will apply to any given cryptosystem. What is of interest to the user of a given cryptosystem is: what kinds of attacks does this cryptosystem protect me against, and what attacks can be successfully mounted that I must be aware of? For example, in almost any instance it is possible to mount a denial of service attack, in the most extreme case by destroying the communications medium. Another possibility is that the entity on the remote end has been corrupted by an adversary, either through social means, say a bribe or a threat, or possibly by the corruption of the operating medium, for example, a virus monitoring data after decryption. Such extremes may be rare, but the example further illustrates the need for understanding exactly what a given cryptosystem provides, and what it does not. In this section some of the attributes of the WTLS protocol are described. In particular, general classes of attacks that are prevented by the protocol are outlined, and some of the things it does not prevent are discussed as well.

As outlined, the WTLS protocol consists of a public-key handshake used to derive shared keys followed by subsequent per-packet operations. Assuming that the per-packet operations use strong algorithms like triple-DES and full-strength SHA-1, then the security attributes of a WTLS-secured connection depend heavily on the key exchange method used in the handshake.

There are three possible key exchange modes used in the WTLS handshake: anonymous, server authenticated, and client-server authenticated. In the first scenario, the handshake uses an anonymous key exchange mode, such as "DH_anon." In this case, each party generates short-term or ephemeral keys for the handshake, and no certificates are exchanged. As such, there is no mechanism to verify the identity of the correspondents. Nonetheless, once a handshake has been established, the use of strong symmetric encryption ensures that sent packets can only be decrypted by the entities who participated in the handshake. The use of MAC algorithms provides assurance that each packet received originated with the entities who participated in the handshake. Such a mode might be appropriate if entity authentication is unwarranted or can be accomplished by other means, perhaps by the use of passwords. However, this mode is vulnerable to the infamous man-in-the-middle attack, whereby an entity between the client and server plays both roles simultaneously. This completely defeats the scheme, since the entity can learn all transmitted information and can change or inject new information. The advantage of this mode is that it does not require the existence of a public-key infrastructure, but it has the disadvantage of being completely defeated by an active adversary.

With server authentication, the handshake uses a key exchange mode, providing authentication of the server to the client. For this to be possible, the server must have a certificate containing its public key signed by an authority known and trusted by the client. Thus, some public-key infrastructure must exist, and in particular a trusted copy of the root certificate must be accessible to clients. After a successful handshake, the client is assured that she is involved in a real-time communication with the server named in the received server certificate. In particular, if the client receives and verifies a certificate containing the identity "Bob's Books," then the client is assured that the entity with whom she is communicating, and to whom she may soon send her credit card information, is indeed Bob (actually, Bob's Web server), and not another entity only claiming to be Bob. No adversary, not even one with complete control over the intervening network, can fool the client into thinking otherwise. Thus, the client can be assured that her credit information is being sent to reputable persons, at least in the case that the client believes Bob is reputable. As with the anonymous key exchange, subsequently sent and received packets cannot be decrypted or generated by parties other than those involved in the handshake. Thus, this scenario provides a reasonable level of security. This type of transaction is commonly used to protect electronic payments on the Internet today.

Careful consideration reveals that the server, the bookseller in our example, has no guarantee about the identity of the client with whom he is communicating. For example, assume an adversary manages to acquire the credit card information of Bill Gates. Then this adversary can connect to Bob's Books, claim to be Bill Gates, send the appropriate credit card information, and there is very little the server (Bob) can do to detect that the entity on the other end of the connection is not Bill Gates. This is a consequence of a handshake protocol where the client is not authenticated to the server. In the case of credit card transactions, the merchant will likely check with the credit card company to ensure that the card has not been reported stolen, so it is possible at least to limit some losses. Other means of authenticating the client can be used, if desired. For example, the client may be required to produce a password that was agreed on at some earlier time, say account setup. Telephone banking has a similar situation in that you call the bank's phone number, so you are reasonably assured you are actually talking to the bank, but the bank has little assurance you are who you claim to be. They often solve this problem by asking questions concerning your mother's maiden name, your account balance, your address, etc. There are valid reasons for having a protocol where only one entity is authenticated. One example is a medical database site, where users would like to know the information is genuine so they want to authenticate the site, but the users themselves would probably like to remain anonymous. Similarly for a stock quote service, clients want to be sure the stock information is genuine, but may not wish others to know which stocks they are interested in, and so prefer to remain anonymous.

The final scenario involves a handshake where both parties are authenticated to each other. In the case of WTLS, this requires the use of a PKI incorporating both servers and clients. Clients must be issued certificates, and must be able to provide a chain of certificates beginning with their own certificate and ending with a certificate trusted (i.e., already possessed by) the server. Conversely, servers must be issued certificates, and must be able to provide a chain of certificates beginning with their own certificate and ending with a certificate trusted (i.e., already possessed by) the client. This scenario provides a strong level of security in that there is no chance that an adversary can convince anyone into thinking they are communicating with someone with whom they are not actually involved in a communication. Such security is desired in many types of transactions, such as stock trading and bank transfers. Applications involving access to company intranets will also require client authentication, so this mode is appropriate and provides considerably more assurance than standard password-based authentication.

7.4.4   Comments on WTLS and WAP security

This section provides some generic comments about the WTLS protocol, and security in general in WAP. It also raises some of the issues worthy of consideration when implementing security in WAP.

WTLS offers a range of cryptographic algorithms for achieving the same fundamental goal. For example, both SHA-1 and MD5 can be used as the basis of MAC algorithms. DES, RC4, and IDEA can be used as base algorithms for providing confidentiality. When choosing the algorithms and parameters, there are a number of issues of which one should be aware. For example, the output of the hash algorithms SHA-1 and MD5 can be truncated to provide weakened MAC algorithms, which provide slightly lower bandwidth at the cost of less security. It is hard to imagine a situation where the data being transmitted are worthy of protection by HMAC and SHA-1, but where the tenuous security offered by an output truncated to 40 bits is adequate. A similar statement applies when considering encryption algorithms and key-exchange modes. In general, the designer of a general purpose cryptographic algorithm strives to make few assumptions about what kind of data the algorithm might protect, and thus attempts to build as many security attributes into the system as possible. A similar situation exists in the deployment of WTLS. It is a fairly low-level module in the WAP network stack, and many higher level applications will depend on it for security. Hence, it is prudent to deploy WTLS implementations having the strongest, most robust cryptographic attributes available. In more practical terms, this means deployed WTLS implementations should strive to use established encryption algorithms with at least 80-bit keys and strong MACs having outputs of 80 bits or more. The use of established algorithms like triple DES and SHA-1 is suggested, the public-key algorithms used should be equally strong, and the use of certificates is encouraged. Thus, from a strength perspective, the use of elliptic curves having 160 bits or more is recommended, as is the use of RSA keys having 1024 bits or more.

The WTLS protocol was designed to work even if packets are dropped or delivered out of order because of the frequency of these events in some wireless networks. While it is fairly easy to reorder out-of-sequence packets, dropped packets are another matter entirely. The WTLS protocol, when operating over a connectionless transport like UDP, will not fail or halt if some packets are dropped; in fact, in some modes this cannot even be detected. Application developers should be aware of this. The implication is that applications should not depend on WTLS to protect the integrity of a long message.

Another minor issue is that some WTLS alert messages can be sent without authentication of origin. Thus, it is possible for an adversary to insert some alert messages. Of course, this is always possible before the handshake is completed, since there is no shared key with which to apply the MAC algorithm. Generally, such attacks amount to denial of service attacks, and thus cannot be prevented anyway, as noted earlier. Nonetheless, it is an item to be aware of.

A slightly technical issue is that WTLS, like TLS and SSL, does not provide nonrepudiation of packets, since all packet protection is done using symmetric key cryptography. Thus, any packet that can be constructed by the client, say, could have been just as easily constructed by the server. Companies deploying applications requiring strong nonrepudiation properties should thus be aware of any legal ramifications of using WTLS in an attempt to prevent repudiation.

Other security components of WAP are in standardization. Of particular interest is the WMLScript crypto library specification, and the WAP WIM specification. WMLScript is a scripting language that allows applets to be downloaded and run on the mobile device. The WMLScript crypto library (see also Chapter 2) defines a mechanism for allowing users to see a text string, and digitally sign the text string. This solves the nonrepudiation issue mentioned above. The WAP WIM specification is concerned with providing hardware tokens, such as the GSM subscriber identity modules (SIM), which hold long-term user keys and can be accessed to do cryptographic operations involving those keys. These specifications are nearing completion. The WAP Forum also continues to address issues associated with implementing a PKI as well as with enabling end-to-end security architectures.

7.5 Conclusions

The WTLS specifications, and other security initiatives in WAP, provide a robust, efficient basis for providing security in WAP applications. It is vital that WAP deployments integrate security now so that they will not be found lacking in the future when integration of security will be considerably more expensive, and so that wireless WAP applications can compete in terms of functionality with traditional wired applications.

Selected Bibliography

[1]   Certicom Corp., *An Introduction to Information Security*, 1998. Document available from http://www.certicom.com.

[2]   Internet Engineering Task Force, *Internet Public Key Infrastructure (PKIX)*. Documents available from http://www.ietf.org.

[3]   Menezes, A. J., P. C. van Oorschot, and S. A. Vanstone, *Handbook of Applied Cryptography*, Boca Raton, FL: CRC Press, 1997.

[4]   Schneier, B., *Applied Cryptography: Protocols, Algorithms, and Source Code in C*, New York: Wiley, 1994.

[5]   Stinson, D., *Cryptography: Theory and Practice*, Boca Raton, FL: CRC Press, 1995.

[6]   WAP Forum, WAP WTLS—Wireless Application Protocol, Wireless Transport Layer Security Specification, February 1999.

CHAPTER

8

Contents

WAP for Operators

Johann Reindl

8.1 Introduction and background

The focus of this chapter is on the introduction of WAP in GSM markets in Europe. Like most wireless markets, GSM markets in Europe have not been formed by free-market forces but through the regulatory regimes of national governments, which are conforming to a large extent to the regulatory guidelines of the European Union. Depending on the country, most national regulators started the GSM market with national licenses of GSM 900. In markets like the United Kingdom or Germany, two operators had a head start before the regulators started to increase the competition by introducing PCS 1800 licenses. These later entrants severely increased the competitive situation. Because of the disadvantages of the late start (e.g., reduced coverage, unknown brands, etc.), they had to decrease prices tremendously to further open the market and attract new customer segments.

The result is that most markets are enjoying tremendous growth in an oligopolistic market structure.

Most GSM markets went through the same market development stages (see Figure 8.1). In the early stage of the market introduction, coverage was the major discriminator, especially for the corporate market, which traditionally is less price sensitive. In stage two, operators started to tap the residential market. They therefore concentrated more on pricing. For the residential market, the decisive price factors are the initial cost of ownership with focus on the terminal/price relationship and the monthly fixed charges. Usage cost is a lower consideration because the first drivers are ownership and passive contactability. Through subsidizing mobile terminals, the initial cost of ownership has been reduced. Later in the development, operators looked at tariffs to increase the usage per customer. In the next stage of the development, where penetration is already very high (see markets in Scandinavia), services provide an increasingly important tool for differentiation and revenue generation. Operators like Radiolinjia in Finland differentiated themselves through new value-added services. Radiolinjia positioned itself as an innovator by marketing SMS messaging and content services. These new applications serve two goals—first, they support the operator's aim for differentiation, and second, they compensate for the slower revenue growth resulting from the continuous price cuts for traditional voice services.

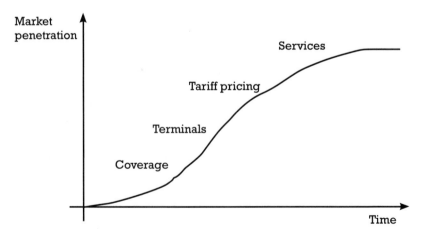

Figure 8.1   GSM market development curve.

The competitive situation has tremendously increased since there are now three to four competitors per country depending on the size or development stage of the country. The competitive focus starts to move away from pricing because the price level in some markets is so low that profit margins are very small. In Austria, for instance, prices for internal mobile network calls were at 1 schilling per minute (ca. 7 Euro cents) [1]. The focus is now moving towards services. First operators offered value-added services like voice mail, SMS, and operator services almost for free, and now they enhance their offerings with content services like traffic, news, weather, sports or stock information. These services have experienced strong growth despite the fact that the marketing focus has been on tariffs. SMS especially is enjoying an almost phenomenal growth, particularly in the young-age market segments. Between 1997 and 1999, the German GSM market experienced around eightfold growth in SMS mobile-originating traffic alone [2].

Operators also use new technologies to trial new services other than voice-based or SMS-based content services. Cell broadcast has been in trials in many countries with limited success. The main hurdle for the mass-market development of these new technologies is the limited usability. The SIM application toolkit (SAT) seems to be a more successful enabler because it offers better usability for mass-market applications. Unfortunately, SAT-enabled handsets and new SIM cards are required. The latter is, in terms of logistics, particularly problematic in mass markets. Even though WAP has been initiated by handset manufacturers, it is the latest of these innovative enabler technologies with very high potential because it leverages on existing technologies.

Although data services have not fulfilled the expectations of GSM operators, there is an overall understanding in the GSM community that data networks like GPRS and UMTS will be the next major innovative phase in mobile markets. To make these investments successful, operators have to introduce new services which take advantage of these new network capabilities.

WAP could be the right tool for the current phase of development because it enables an IP type of services with current network functionality. The value statement is along the lines of "more than voice" or "the mobile becomes your infoterminal." Data services require a learning curve for operators as well as for the market. Most likely, it will take more time to develop the market to use a mobile phone for nonvoice services than for telephony services. The market education with low-bandwidth mobile data services seems an ideal preparation for a successful introduction of high-bandwidth services.

The core question of this chapter is: Do operators benefit from WAP, and if so, how can they best take advantage of this new technology to contribute to their business objectives? In order to answer this question, operators' needs have to be identified first. Having done that, one has to look at the market situation in order to decide if, and how, WAP can be used to fulfill operators' needs.

8.2   Operator needs

As described in the market overview, GSM operators are by and large confronted with a new market situation. The main problems they face are:

▶ Strong competition on price;

▶ Falling margins on mobile telephony services;

▶ Low degree of differentiation with traditional means like coverage, terminal, and tariff pricing;

▶ Ongoing high churn rates of existing customer base.

Most operators are therefore trying to focus on new services and branding in this new market development phase. First of all, they need to identify new means of differentiation to retain their existing customers and to win new customers. Since price levels are quite low, it is difficult to achieve a market impact by minor incremental price decreases. In addition, falling margins will hurt operators' profitability once subscriber growth starts to decrease.

Second, revenue growth will be difficult in the light of falling tariffs and decreasing subscriber growth. Although there is some kind of price elasticity on usage, it cannot compensate for stagnating or even falling revenues per customer. It is therefore important to offer new high-value services which can compensate for the decrease in telephony services.

From the introduction of data services, for instance, operators can benefit twofold: first, on the bearer level from the increasing traffic in the network and therefore better profitability by gaining higher capacity utilization, and second, from new revenue streams from the application level. Many GSM operators are concentrating on introducing HSCSD or GPRS services. The critical question, however, is if there are data services which generate enough revenues to make these infrastructure investments profitable. This is particularly relevant in the light of the historical experience with data services.

The market development of data services so far has been rather disappointing. In Germany, for instance, up to 1998, only 2% to 3% of mobile customers used mobile data services [3]. Only SMS messaging, which recently enjoyed tremendous growth, is an exception. This is due to the take-up of young market segments which are both price sensitive and very open to new technology and media. Apart from this niche, data services are established in vertical markets like dispatch services or in horizontal markets like corporate intranet access for mobile workers or executives who are on the move and use their laptops and mobile phones to get updates on work situations. Both of these markets show limited growth, the former because in vertical markets, mobile applications have to be adapted to whole work-flow management applications, which are very company specific and resource intensive to implement. Efficiency gains are therefore difficult to realize. Moreover, the total solution is still very cumbersome—you need a mobile phone with data capability, a laptop, the right software, access to the corporate intranet, and often a PC Card. Apart from that, with 9.6 Kbps it is not fast enough to use normal Internet or intranet applications. As Ovum has identified in a corporate study, most corporate respondents see security, data rates, and network quality as the main barriers to take up mobile data services. This might be one of the reasons why e-mail is one of the most popular mobile data applications today [4].

WAP has the capability to offer real mobile intranet and Internet applications with marginal cost. Therefore, it has the potential to truly reach the mass market. For operators, WAP first would increase the often low-capacity utilization of CSD in GSM networks, and second, develop the market for GPRS and UMTS networks. Both networks require such heavy investments from the operators that they can only be profitable for the mass market. This is particularly important when looking at the timetables. The first GPRS networks are planned for commercial introduction in mid-2000 and UMTS networks could start as early as 2002. Operators do not have much time to go through a learning and market education period. Rather, they have to be ready once new technology becomes available and take every opportunity to teach themselves and educate the market.

The rather unsuccessful provision of dedicated data networks showed that networks with mostly vertical services cannot be profitable in competition to GSM. The market pull for high bandwidth can be best generated through successful low-bandwidth applications. For the corporate market, operators can offer IP access through WAP, which allows for a single point of access for both fixed and mobile users. Therefore, the intranet access does not have to be changed once GPRS-based IP networks come. The fact that IP traffic has already overtaken voice traffic in the fixed network shows that data applications will dominate telephony services. Since wireless can be seen as a value-added extension to wireline, it is only a matter of time once the same development takes place in mobile markets. The missing link currently is on the service and terminal side.

Bearing in mind that handset manufacturers will introduce WAP-enabled phones to boost their sales, there are no reasons why operators should not introduce WAP, especially in the light of the competitive situation in most markets. Following the press coverage of Telecoms '99 in Geneva, the mobile industry as a whole seems to take up WAP. The opportunities for operators by far outweigh the risk of an unprofitable investment, especially in comparison with GPRS and UMTS types of investments. WAP could also be a stepping stone for a successful introduction of data services to the mass market, which will be absolutely necessary to make GPRS and UMTS investments profitable.

Having decided for WAP, the question is how to introduce WAP. In other words, what type of services and in what role? To answer the latter part of the question, WAP business models have to be discussed, which will be done at the end of this chapter.

8.3 Customer requirements

When introducing WAP, operators should concentrate not on the technology, but on services and user benefits. Having identified those, one can start to design business models and to judge the value of WAP.

As with most services, customer requirements reflect customer expectations. Therefore, the positioning of WAP services will strongly influence customer expectations. If the services are positioned as VAS (value-added services), users will compare them to existing services they already use (e.g., SMS or voice-based services). In the case that WAP services are launched as mobile Internet services, users will compare them to Internet services in terms of usability, functionality, price, etc. Positioning is a critical success factor depending on the target group and the type of services. Of course, positioning also heavily depends on the competitive market situation. Depending on the market power, operators have to follow or at least take into account where competitors and the overall market are going.

As with most new technologies, it is difficult from a customer's perspective to imagine what Internet access from a normal mobile phone will be like and what services he or she would prefer to use. Therefore, customer requirements are derived from other mobile and Internet applications and are only applicable to a limited extent. Nevertheless, the results can be used in designing the service offering and marketing strategy.

In order to design a successful service offering, basic user barriers have to be taken into account. Psychological user barriers can influence the attitude towards new technology. People who have limited or no experience with the underlying technologies (Internet and mobile communications) are less likely to adopt WAP services than people with extensive experience. Technological user barriers exist in terms of the provision of the right hardware components, quality of service, service usability (response time, speed of transactions), and, depending on the type of service, security requirements (see Chapter 7). Some of these criteria are critical for the success of WAP services. The history of mobile data services is a classic example of not fulfilling these requirements. The requirement of additional hardware equipment reduces the user group due to inconvenience and cost. The same is true for accessing the Internet with a laptop and mobile phone. Bandwidth, pricing, and equipment requirements are not fit for the mass market. Cell broadcast is an example of the importance of usability and standardization. The difficult service access and phone-specific behavior decrease user acceptance. Both the handset manufacturers and operators play a critical role in making the total service offering usable. Mobile users are used to extended battery life, small terminal size and weight, and low terminal costs. If these expectations are not fulfilled anymore, it is unlikely that a large number of mobile users will adopt these new services. But data services have additional requirements. One- or two-line display sizes are obviously not suitable for text-based services. From the operator side, speed (access and response time) is as important as security and convenient user access (subscription, browser configuration, and service access have to be convenient and low cost has to be suitable for the mass market).

For the successful launch of WAP services, it is very important to target the right user group. It seems important to first target the so-called lead users or innovators, who animate followers through their ownership and usage. Applying such a targeted marketing strategy is most effective with new technology products. To identify such high potentials, it is necessary to identify the key characteristics of these users. To just target people who are open to new technology could be risky since they do not have the relevant experience with similar products. It is therefore safer to identify users who have experience with related products or services. The affinity to WAP services is influenced by both the experience with mobile phones and services, and the usage of already existing Internet or on-line services. To further differentiate within the mobile experience, the usage of data services like SMS indicates an even higher degree of affinity.

Since WAP is the combination of mobile and Internet services, the target group or innovators will already have adopted both mobile communication and Internet services. As market research figures show, there is already a large overlap between mobile and Internet/on-line users. In Germany, recent research shows that in 1999, on average more than one-third of mobile customers used Internet/on-line services regularly [5]. From this one can draw the conclusion that WAP will be less a tool to further extend the mobile market by attracting new users, but more a tool to increase usage and revenue with existing customers by enlarging the service portfolio.

Having identified the potential user group, the next step is to identify the preferences of these users to design the right service offering. The preferences for the service offering can be derived from primary market research as well as from the usage pattern of Internet/on-line users and, to a limited extent, the usage pattern of SMS-based content and messaging services.

Nokia undertook a primary market research study [6] on the demand of mobile value-added services. Table 8.1 illustrates the results on the services in order of the favored demand. The percentage presented is the cumulative percentage containing "would use" and "would be likely to use" answers. No matter how detailed one looked into the results, banking services always scored the most points. Even comparing different interest-level groups, the demand for banking services is always highest. In addition to banking services, phonebook and e-mail types of services seem to be very successful. The services that generally face poor demand are entertaining type of services like jokes, biorhythms, and sunrises.

Table 8.1
Demand of Mobile VAS [6]

Services	Level of interest (%)
Banking	93.3
Phonebook	88.9
E-mail	84.4
Ringtones	77.8
Dictionary	77.8
Remote control	77.8
City navigator	75.6
Weather	75.6
Pizza order	66.7
Traveling businessperson	62.2
News items	55.6
Big events	44.4
Gambling	42.2
Stock info	40
Sports	37.8
Stock portfolio	35.6
Jokes	26.7
Free time	24.4
Sunrise	20.0
Biorhythm	11.1

What type of service attributes represent the highest utility value to the end users? A further study of the characteristics and attributes of the services has been done in order to derive marketing implications. The results are presented in Table 8.2. For instance, 89% of the showed sample say that banking services are personal, whereas 89% report joke services belong merely to free time.

From these results three main services categories can be derived [6]:

▸ Entertainment services (information about events, gossip, local news, culture, etc.);

▸ Infotainment services (database queries like job search, news about economics or politics, etc.);

Table 8.2
Service Attributes [6]

Banking services	Attribute importance (%)
Personal	89
Useful	86
Use both for work and free time	56
Pragmatic	83
Joke services	**Attribute importance (%)**
Mass service	58
Useless	69
Free-time usage	89
Entertaining	99
News	**Attribute importance (%)**
Customizable	41
Useful	73
Use both for work and free time	55
Pragmatic	83

▸ Transactional services (e-mail, banking, on-line games, travel booking, shopping, etc.).

Entertainment services show the lowest overall demand and can be interpreted as being mass services. Their function is only to provide entertainment for free-time usage. There are, of course, user segments which might rate them differently.

Infotainment services are facing a moderate level of demand and are seen quite opposite from entertainment services. They are seen as pragmatic and quite useful. The distinction between mass and personal services becomes blurred. Here, the source of interest is the personal touch of the content. For instance, the stock quote of one's portfolio or the weather information of one's city or holiday resort can be quite personal. Information services are used for both work and free time.

Transactional services generate the highest level of interest. They are often used to perform routine tasks and show a high degree of personalization. This might be the reason why transactional services seem to have a higher utility value for the user. It has to be mentioned here that these services also require a larger technical effort and are therefore more suitable for a later phase of the WAP service introduction. For instance, banking or e-commerce requires end-to-end security. The necessary functionality is currently not standardized by the WAP Forum.

8.4 Critical success factors for WAP service introduction

Despite a number of WAP-specific critical success factors (CSFs), a generic list of CSFs for service introduction also has to be taken into account.

8.4.1 Generic critical success factors

As with most services, pricing is also critical for WAP services. From a user's point of view, the total service offering has to be priced appropriately. As handsets play an important role, the terminal pricing has to follow the normal price categories. The mass market is not willing to pay a premium for the WAP functionality. The user values the phone in total; WAP is only an additional feature. Ideally, WAP browsers come with the new generation of phones so that WAP is an additional reason to buy the phones. Since the phone market is very image driven, users are not willing to accept limitations on size, battery life, design, etc., just for the additional browser software. In addition, particularly younger age groups tend to be brand loyal and will therefore not switch for the sake of WAP alone.

The other important part of the pricing decision is concerned about service, access, and applications—from a user's perspective they are often the same. On top of the already applied pricing options like subscriptions, pay-per-usage time, and pay per event, it is also possible to charge the access separately from the application as well as to charge for data volume. Depending on the launch strategy, the pricing decision should be along the lines of the Internet pricing model. It is, however, questionable how much users are willing to pay for content services, even with a degree of personalization, bearing in mind that most content is free of charge on the Internet. If planning to charge for content, it is very important to educate the market right from the beginning. Starting with a free-of-charge model will make it very difficult to introduce application charges later on.

Since WAP is a new standard and WAP terminals and platforms became available at the end of 1999, operators have to decide on the timing of the launch. If they follow a leadership role, they can position themselves as innovators and are free to decide on how to position WAP services. This implies that they have to develop the market with all the usual development costs. If they choose a follower position and wait to see what other competitors are doing, they more or less have to follow the competitors' positioning to avoid confusing the market or bearing the cost of reeducating the market. Depending on the market situation, the latter might be an impossible strategy.

The positioning of WAP services is key to a successful launch. Depending on the timing and the target group, the positioning strategy has to be developed. Market communication, the naming, and the value statement are critical in managing customer expectations and influencing consumer demand. Packaging is also critical—is WAP only an additional feature for a mobile phone, or is WAP the reason for purchasing the phone? The latter statement might be too early for the current development phase of the mobile market.

The choice of the right sales and distribution channels is critical in reaching the right target group and getting across the more complicated service offering. In order to reach the mass market, a critical number of Points of Sale (POS) have to be accessed. Alternative channels like the Internet should also be taken into account.

Like with all mass-market products, WAP services have to comply to usability and sellability requirements of mass markets. The browser configuration or the service subscription processes have to be automatic and not require time or effort; in other words, the cost for the sales channel or the user has to be low. If the salesperson needs additional time or effort to convince users of the benefits of WAP, sales incentives have to be provided. Experience also shows that the response to customer self-activation is very low even if it is associated with freebies or vouchers. Preconfiguration and automatic service activation are critical for success. An automatic service representation on the mobile terminal is also critical for service adoption. Especially at the beginning of mobile usage, consumers play with the phone and will activate or use services if they are represented as a hot key or menu point. A positive example is SAT services, whereas cell broadcast and SMS services are rather on the negative side. The SAT standard defines a menu point or softkey as part of the mobile phone menu and therefore leads the user to the service.

A large number of value-added voice services are not used because customers are not aware of them. Users simply do not remember the phone numbers or the complicated syntax of SMS-based value-added services.

8.4.2 WAP-specific critical success factors

Similar to the Internet, a range of free WML content will be available in the long run. Established content providers will pick up the new standard to adapt their offerings to also reach the fast-growing mobile community. But in the initial phase, there will be a very limited offer of WML-based content, especially in national languages other than English. Operators will therefore play an important role in creating a critical mass of content services or other applications so that the end user can experience the value of WAP. This is a prerequisite to get the necessary pull effect for content providers to adapt their services.

In order to get services to the requested quality, the functionality in terms of security, personalization, and actuality has to be provided. It is important to bear in mind that mobile users have by experience higher expectations in terms of quality of service than traditional Internet users. Availability, speed, and quality are critical to attract mobile users who are not familiar with the Internet. The management of expectations seems to be critical in positioning WAP services.

The availability of a large enough portfolio of terminals is also critical to the adoption of WAP. Interoperability of terminals, services, and WAP gateways is an absolute must for a successful market introduction. Since WAP is focused on mass-market mobile terminals, the availability of inexpensive, easy to use handsets is critical to the success. As the current market experience shows, a high number of mobile users show strong loyalty to terminal types and brands. It is therefore unlikely that mass-market users will change their terminals because they want to use WAP services. In a lot of countries in Europe, terminals are heavily subsidized by operators. The value of an offering is therefore strongly dependent on the type of handset as part of the packages. In markets like Germany or the United Kingdom, where subsidies are quite high, it takes about 2½ years to churn a total terminal generation [7]. This in turn is a positive aspect for the development of WAP, because the distribution of WAP phones can be strongly influenced by the operator.

For the high-end users who already use the PDA type of terminals, it is questionable if WAP is really taking off, bearing in mind that with new bearer services like GPRS, due to be launched in mid-2000, standard Internet and intranet services can be accessed in a comfortable and cost-effective way with normal HTML browsers.

8.5 WAP services

The following WAP services are categorized from a network perspective. Three main categories have been differentiated: Internet-based services, intranet services, and infranet services. The latter stands for the operator's infrastructure, where VAS platforms or customer self-care applications are located.

Internet/content services are horizontal applications (i.e., they are suitable for both the corporate and residential markets). They can be further classified in terms of usage type:

▸ Infotainment services, as the word says, encompass both information services (e.g., database queries or content push services) and entertainment services like games.

▸ Transactional services cover real-business transactions with a real obligation like banking, brokerage, and e-commerce.

▸ Messaging services are communication-based services like e-mail and chat.

To evaluate their relevance for WAP, indicators, which evaluate the utility value for the user, should be defined. Since the early adopters of WAP applications will very likely have Internet/on-line experience, the usage pattern of Internet services can be used as a benchmark. E-mail, banking, and infotainment services are at the top of the list in most surveys. For the mobile user, the e-mail functionality could be combined with fax or unified messaging features (e.g., text to speech) so that mail can be printed at a nearby fax or can be listened to directly from the phone. Banking applications are more difficult to implement because of augmented security requirements. As a first step, bank statements could be accessed and real transactions can follow once end-to-end security is available. In the next step this basic demand pattern has to be enhanced with mobile-specific value criteria.

Mobile users put a high value on reachability anywhere at any time. This can be directly applied to the messaging type of services because of the similarity to telephony services. For the other service categories, the anywhere and anytime translate to actuality because having the right information anywhere at any time means having the most actual information. Push services like stock quotes, traffic information, changes to flight/train schedules, and last-minute offers deliver a high value to the user. The more actual the information, the higher the utility value.

Besides actuality, mobile communications can support mobility by providing relevant information for being on the move. Within infotainment services there is a category of content which is particularly (almost only) relevant for mobile users. For instance, traffic information, route planning, town navigation, points of interest, or city guide applications deliver the highest value if people are lost and looking for help. In order to increase the usability, the application should get the user location information from the network or GPS in order to provide the most relevant data.

As a general development road map, content services can be often enhanced with real transactions. For instance, a film guide or train schedule database can be enhanced with reservation or booking functionality. Since transactional applications are much more difficult to implement, it is better to offer basic services first in order to decide on further development depending on user demand. Because a lot of transactions are currently done by telephony, it is already possible to use the WTAI feature to access transaction or booking agents via voice directly from WAP applications (see Chapter 4).

As with existing Internet/content services, operators cannot cover the whole value creation process since they do not have either the competence or the resources. It is therefore even more important to identify the role operators want to play in this category. Although this question will be further discussed at the end of this chapter, the real added value where operators do have a distinct competitive advantage for these types of services can be reduced to the following five points:

- Knowledge of customers and identification/authentication through SIM cards, valuable for transaction-based services and personalization;

- Location information in different quality levels, valuable for personalization and convenience;

‣ Mass-market capable billing systems, valuable for e-commerce;

‣ GSM network security, valuable for transaction-based services;

‣ Face to the customer with operator logo and portal through handset subsidies and configuration.

Intranet services are targeted at corporate mobile users who want to access their intranet while being on the move. Two main categories are classified here: horizontal applications and vertical applications.

Horizontal applications include applications like corporate e-mail, corporate directory, organizer, and access to corporate groupware. In these applications, it is the classic tradeoff between centrally held applications with mobile clients or local applications in the mobile terminals. Since connectivity becomes more important, the client/server approach increases the potential for WAP.

Vertical applications encompass all sector-specific applications like work flow management for dispatch services or service management organizations. Most of these applications are running on the corporate or client side because they are mostly connected to on-line systems. Synchronizing and updating centrally held databases is very often the key benefit to the user. The operator's real added value for intranet customers is a secure and comfortable access from the mobile network to the intranet. For instance, user identification and authentication for intranet access can be carried out by the operator. For smaller corporate customers with no real intranet, there is an opportunity for centrally hosted applications with the associated user data. The investment for small companies to use these services is minimal. The development of VPNs in telephony shows the potential for data type of VPNs. The border between secure Internet and intranet becomes indistinguishable.

Intranet services or customer self-care type of applications include operator-specific services like the control of other services (e.g., call management, mailbox, messaging, the activation or cancellation of services with mandatory subscription, tariff management (i.e., information about tariffs and the change of tariff plans), access to call histories or electronic bills, and any other type of contract or service handling). From the operator's perspective, the motivation is clearly to gain efficiency and to increase customer satisfaction because instantaneous access to the requested information is difficult to satisfy through the normal customer-care systems. This becomes even more relevant in a situation where the number of subscribers and services is growing fast. Critical for success is a fast and easy navigation to the requested information or action, otherwise customers will not switch from already familiar and easy to use channels. This seems to be the real challenge in this category. Operators should think carefully about applicable applications for this media. Visual management of complicated services like unified messaging or organizers adds value to the currently available voice-based management functionality. However, the complicated navigation to particular roaming prices has to be carefully thought through. Often customers need incentives either in the form of cost savings or increased convenience to use such services.

In comparison, WAP seems to be capable of reaching two goals: to migrate SMS users to the higher value WAP services, and to access a larger part of the traditional voice users. Compared to SAT services, the operator does not have to design a new process for exchanging SIM cards but relies on the existing market mechanisms for exchanging mobile phones in markets with terminal subsidies. There, mobile phones get exchanged about every two years. The potential user group therefore grows naturally. Since WAP is based on existing Internet technology, it is very likely that the number of WAP services will grow in line with the number of WAP users and thereby increase the utility value of WAP without necessary investment from the operator side.

However, since WAP is an open standard and the service offering will rely to a large extent on the Internet, it will be difficult for operators to create a sustaining competitive advantage. Only in the initial market development phase can operators differentiate themselves with a particularly broad, high-value service portfolio. The Internet already demonstrated that on-line providers like T-Online or AOL in Germany had to differentiate themselves with services like premium content, e-mail, and banking, and therefore are able to demand higher prices. The question will be how the WAP market will develop and what role mobile operators seek to take—pure access providers like IAPs in the Internet, the extended IAPs like mobile ISPs or even the role as mobile on-line community providers.

Independent of the role in the value constellation of the WAP market, operators will even benefit from the reduced role as pure mobile IAP from increasing revenue and higher utilization of the data network capacity. Especially as telephony prices have fallen tremendously under severe competition, it is possible to skim off new services like WAP. Depending on the market segment, a mixture of monthly and usage-based fees can be a successful way of capitalizing on this new market.

Even in a later phase when WAP usage prices will fall to the low level of telephony prices, the added usage will create additional revenues for operators.

If mobile operators are offering WAP services through their own proxy gateway with a terminal configuration process, they already offer mobile Internet access. Additional services like e-mail are often offered by operators as they migrate their voice mail systems to unified messaging systems. They only have to bolt the WAP functionality to some of their services and the value added of an ISP is provided. Personal WML sites can be offered in a later stage once easy to use service creation tools are on the market. Since WAP is currently only text based, the service hosting will not cost a lot of storage. The last value of content or content portal provision can be offered by operators on their own, but it seems more reasonable to cooperate with established players in the Internet to adapt their content portals to WML. However, mobile operators have a strong strategic position to preconfigure the handset browsers to their home portals and therefore sell portal space to established content or Internet portal players.

This service category is an absolute must for operators when introducing WAP. Based on that, operators can decide how far they want to extend their role into Internet and intranet type of services. The following section will discuss the pros and cons associated with that question.

8.6 WAP business models from an operator's perspective

Since WAP can be seen as a wireless version or extension of the Internet, it is by definition an open business model. However, operators have the option to apply the WAP technology in a closed domain, similar to the application of Internet technology in intranet domains. In the following, this case is called the closed model. In the closed model, operators use their own applications or the application of a trusted third party to offer them to their own customers. There would be no option to access the free Internet. From an operator's point of view, this option is using new technology to enhance the capabilities of existing value-added data services like SMS. On the positive side, operators can control the service domain and are therefore in a good position to ensure a high quality of service with a high level of security. Because the operator has control over the service domain, it creates the highest possible value within the value chain. However, the time, resources, and competence required to develop a high-value service portfolio should not be underestimated. The decision for such a strategy should be taken in light of the competitive situation and the operator's strategic direction. The closed model might be suitable for a portfolio of premium content services enhanced with e-mail, organizer, and customer self-administration services. The closed model can be compared to the role of an on-line service provider who is not offering access to the free Internet. This model can be extended to the role of a community provider.

How successful will such a model be in light of the development in the Internet? The analogy to the early days of on-line services like T-Online, Compuserve, and AOL can be drawn when they mainly offered their own premium content services and customers had to pay a high fixed price. A premium had to be paid to access the free Internet. As the market development shows, the lesson has been learned that users want the open Internet model. Hence, most on-line service providers today offer free or inexpensive Internet access on top of their own content services. As an example, T-Online of Germany is positioning itself as an access and portal provider, which means that Internet users from other ISPs can preconfigure it as a portal and its own customers can bypass it to preconfigure other portal sites. If WAP is following the Internet market development, the closed model's sustainability has to be questioned.

Like Internet service providers (ISPs), operators have the option to run an open model with full access to the free Web. Depending on the role the operator wants to play, it can concentrate purely on the access role (analogue to Internet access provider (IAP)), providing a gateway for all wireless users (its own customers, competitors, and international visitors) to the Internet and third-party services. Its own customers would possibly have enhanced services by accessing the operator's own services (e.g., self-care type of services). This reduced role can be extended by offering a broader service portfolio or by providing a portal to premium services. In this case the operator would play the role of an IAP with a content or service portal like Yahoo! or AltaVista in the Internet today.

Table 8.3 provides an overview of the described business models with a rough evaluation from both the user's and operator's perspectives. Although the open model does not seem to have a much higher score than the closed model, the Internet success shows that service portfolio and value of the offering are much more important from a user's perspective than the ease of use and transparency. The lack of transparency and ease of use in the Internet created new business opportunities for content

Table 8.3

WAP Business Models

Business models		Closed model	Open model
Customer focus	Service portfolio	—	++
	Ease of use	+	-
	Pricing/service transparency	+	—
	Value of offering	-	++
Operator focus	Potential for differentiation	+	0
	Control over offering	++	—
	Market potential	-	+
	Development/administrative cost*	—	++
	Potential for customer retention	+	-

* Evaluation is reverse to costs

++ = very high   + = high   0 = middle   - = low   —— = very low

or application packagers or portals like Yahoo!. If WAP is imitating the Internet, it is very likely that the Internet/open model will be adopted.

Depending on the objectives and the strategy of the operator, the closed model might be favorable to the open model. Since differentiation is often a key objective in the launch of innovative services, the exclusivity of the offering might be the right strategy for the initial stage of the service introduction.

Assuming that most operators which introduce WAP will apply the open model, the question about the role an operator should play in the value constellation of Internet-based WAP services has to be answered.

To begin with, operators should carefully analyze the competitive situation they are faced with before moving into new business fields like the aggregation and provision of content or e-commerce services. Since the Internet will not be newly defined by WAP but rather extended, established Internet players will most likely extend their domination to the mobile arena. Mobile operators therefore have to identify their natural role and the areas of real value creation where they can create a strong competitive position. From such a strong base, they can selectively try to reach further into the value creation process by partnerships, acquisitions, or natural in-house expansion.

How should operators identify their "natural role" in this new business field? As has been already pointed out under the previous point of WAP services, mobile users will always have to cope with limitations compared to the fixed-line Internet/intranet. Using the limitations of the mobile world and of the existing Internet, operators can build a strong case for value creation by breaching these gaps. The following lists the most relevant issues:

▸ Because of the mobile-user scenarios and the limited terminal capability, user input should be reduced as much as possible. Personalization is therefore much more important than in the normal Internet. Operators have not only the standard GSM authentication via SIM-card, but also the location information. This can be used in a whole range of value-added applications either in-house or transferred to external partners.

▸ Security is one of the major stumbling blocks for transaction-based services, which create the highest perceived value to users. The current WAP standard does not offer any end-to-end security mechanisms. Operators should try to influence the standardization process so that the SIM-card will be part of any public-key infrastructure (PKI) by hosting keys or electronic signatures.

▸ The mass-market type of billing or payment mechanisms also does not exist in the Internet, but with current operators even on the micro-payment level. Therefore, operators can extend enabler systems like billing and customer care to cover third-party application providers. This might be particularly important for WAP content providers because it is unrealistic to apply the advertising model of subsidizing content to the WAP environment. Micro-billing may be a prerequisite for some providers to offer content on WML.

▸ Since operators control or at least influence the distribution and configuration of handsets, they are in an ideal position to offer a consolidated portal of mobile-specific intranet services and third-party Internet/content applications. This combination creates the strongest position of any mobile portal provider.

8.7 Conclusion

Although mobile data services have so far been rather disappointing, the recent success of SMS services shows that there is real potential. Bearing in mind that upcoming technologies like WAP and GPRS are compensating for existing limitations, there are no reasons why the Internet success story cannot be transferred to mobile markets.

From an operator's perspective, mobile Internet services seem to be an opportunity to compensate for price cuts in telephony services, especially once the subscriber growth starts to slow down. Besides, the currently earmarked investments in the high-bandwidth data networks GPRS and UMTS need nothing more than demand-for-data services. The current demand for the Internet outweighs any predictions, although it is still behind the growth of the mobile market. Mobile Internet could therefore prolong the revenue growth for mobile operators.

In total it is fair to say that the opportunities of WAP by far outweigh the risks for operators. As the successful development of the Internet type of services in the wireline network shows, there are no reasons why a similar development should not happen in the wireless market. The fact that two fast-growing markets, mobile telephony and the Internet, converge almost promises market success. However, as has been pointed out in this chapter, the mistakes from other service introductions have to be taken into account to successfully launch WAP services. Unfortunately, the WAP standard in itself has, at the time of writing, not fully taken care of all the processes and functionalities required to introduce mass-market services. Therefore, operators, applications developers, and handset manufacturers have to keep working together to make WAP a success.

References

[1]   Mobilkom Austria, Web site: http://www.mobilkom, 1999.

[2]   Der Mobilfunk-Report 1999, Plica Marktforschung Analyse, München 1999, p. 116.

[3]   Der Mobilfunk-Report 1999, Plica Marktforschung Analyse, München 1999, p. 113.

[4]   *Mobile Internet and Intranets: The Road Ahead for Corporate Applications*, Ovum Ltd., February 1998, p. 32.

[5]   *T-Mobil Market Research*, 1999 (confidential).

[6]   Ojala, T., 1998, *The Demand for the Mobile Value-Added Services in Finland—Case Nokia Artus Messaging Platform*, Master's thesis.

[7]   Der Mobilfunk-Report 1999, Plica Marktforschung Analyse, München 1999, p. 125.

CHAPTER

9

Contents

Extending Integrated Unified Messaging Solutions Using WAP

Kai Kauto, Ilkka Teppo, and Jarno Haikonen

9.1 Introduction

In recent years the amount of messages people receive has increased rapidly. There exist today new forms of messaging and new devices to handle messages. Behind this development is the explosive growth of the Internet and mobile communications, which have generated new and more effective message forms. The convergence of these fast-growing technologies will even further increase the flow of messages, as Internet messaging will be part of mobile communications. WAP, as the standard linking the mobile communications and Internet worlds, will play an essential role in this development.

This chapter describes how the unified messaging concept achieves the integration of messages and the different types of message formats and how the unified messaging functionality is extended using WAP. It also briefly describes some of the current messaging systems, their characteristics, and how they are implemented using WAP.

The first part of this chapter introduces the unified messaging concept and outlines the effects of WAP on the unified messaging concept. The second part describes the functionality of some unified messaging WAP services from the wireless network operators' perspective. The service concepts and the technology behind the services are illustrated as well. The last part introduces unified messaging WAP solutions in the corporate sector.

9.2 Unified messaging concept

The strong growth of voice mail and e-mail messaging has taken the number of delivered messages to a totally new level. Due to the increased amount of messages and because each type of message may be of a different format and be delivered to separate locations, dealing with this message flow is becoming increasingly time-consuming. Many business users receive a large number of e-mails, voice mails, and faxes per day. Not only the amount of messages creates the message chaos, but also the fact that different message types are accessed from different locations using a variety of devices. E-mails are often retrieved from the e-mail server using a PC, voice mails are accessed from the voice mail system using a phone, and faxes arrive on and can be read from the fax machine. The situation becomes even worse, when traveling, often only carrying a mobile phone. Access to e-mails and faxes becomes difficult or even impossible in that situation.

The unified messaging concept is the solution that aims to allow one to manage the increased message flow consisting of different message formats. The idea of the unified messaging solution is to bring all message types, voice, fax, and e-mail messages, under the same interface; in other words, to make a unified mailbox. Another aim of the unified messaging concept is to implement conversions between different message types to allow access of different message formats from different types of devices. E-mail conversion to voice or fax messages is the most used message conversion service. The overall target of unified messaging is to enable the user to access the unified mailbox and manage all message types from anywhere, anytime, and using any device. Figure 9.1 illustrates the unified messaging concept.

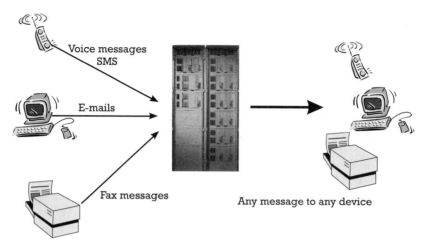

Figure 9.1 Unified messaging concept.

9.3 Unified messaging and the wireless application protocol

Most telecommunication operators' unified messaging concepts are developed on top of a voice mail system. This is the main reason why the interaction with the unified mailbox is traditionally handled using the keypad of the telephonic user interface (TUI). More advanced solutions also have a Web-based graphical interface to unified messaging services. WAP brings the best sides of these interfaces together: the mobility of the mobile phone and the visual interface of the Web. WAP is the solution for the biggest problem in the unified messaging concept: the difficulty in the usage of services from the mobile phone (see Chapter 3 for a detailed treatment of WAP-related usability issues).

The most important aspect of the WAP, together with the unified messaging, is the possibility of having a visual interface to the user's messages. For the mobile phone the visual interface is even more important than for the regular fixed-line phone. In the mobile phone, the keypad and loudspeaker are colocated, and listening and selecting items from a menu structure are difficult to manage at the same time. WAP provides the message management in visible format, which allows an easy management of a large number of messages.

With the unified messaging WAP access, people wanting to communicate are able to do so using their mobile phones with the messaging media that is most suitable for them at any particular moment. Voice, fax, and e-mail messages are available to subscribers regardless of the time and location.

9.4   Architecture of wireless application protocol access to unified messaging system

WAP access to the unified messaging system in the wireless operator environment is built on an application server that acts as a protocol converter between WAP infrastructure and Internet framework as illustrated in Figure 9.2. Towards the WAP framework (in most cases a WAP gateway—see Chapter 5), the application server acts as a Web server providing HTTP protocol and WML content. Towards the Internet framework, POP3 (post office protocol), IMAP4 (interactive mail access protocol), LDAP (lightweight directory access protocol), and proprietary messaging protocols are used. The WAP gateway is physically located in the operator or company's premises.

The application server hosts the service logic and deals with authentication, billing, personalization, and security issues. The same application server usually also handles the Web access to the unified messaging services. The functionality of the unified messaging services accessed using

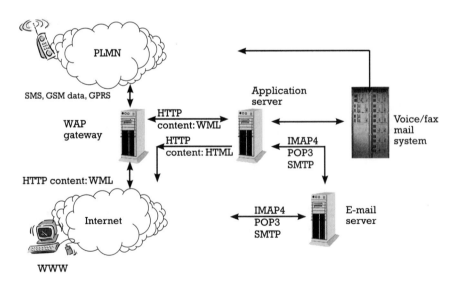

Figure 9.2   Unified messaging system with WAP access.

WAP devices can be the same as from the Web access, because the service logic is hidden under the content represented as WML or HTML pages.

9.5 Using the wireless application protocol to expand unified messaging services

E-mail messaging is the most used service in the Internet. It is fair to say that messaging will also be the most used service in the WAP environment. On the Internet, free e-mail services are increasingly used to attract new and keep existing customers visiting Web sites. It can even be said that in order to own the customer, you have to own the customer's messages. In the WAP environment unified messaging is very likely to be the most effective way to attract and keep customers.

After having obtained customers who are using unified messaging, the unified messaging service portfolio is easy to expand to new areas, including new services or interfaces, by using the same unified messaging interface. WAP will play the key role in expanding the unified messaging service range into new fields. It can be used to expand the unified messaging service possibilities by exploiting the combination of the mobile phone and the Internet.

One way to expand unified messaging is to bring the unified messaging towards, or integrate with, a personal communications portal, which combines messaging, directory, call control, address book, calendar, information, and entertainment services. By combining a PC's keypad and the large screen in the Web access with the mobility of the WAP phone, the best and most compelling services can be created. Good examples of the combination of Web and WAP access to create advanced services are WAP portal services discussed in the next section and the address book services presented in the last section of this chapter.

9.5.1 The WAP portal

The WAP portal service is a good example of a service that makes use of the Web advantages to create innovative WAP services. The portal service enables the subscriber to select WAP services and write HTTP links to WAP content from within the interface of the subscriber's unified messaging Web access. The set of selected links, the user's personal WAP portal, can then be used and accessed using his or her WAP device.

The collection of the selected WAP services and links are presented to the subscriber when the subscriber is accessing the unified messaging services through WAP. The WAP portal service enables the user to write and modify links to his or her most interesting or useful WAP services anywhere on the Internet using the keypad of a PC. It thus provides easy and fast access to WAP services wanted by the subscriber when on the move.

9.6 Wireless application protocol access to unified messaging services

This section introduces the basic unified messaging services from a WAP point of view as well as the functionality behind the services. The services are described from the operator's perspective. The corporate perspective is introduced in Section 9.7.

9.6.1 Accessing voice and fax messages

The key feature of the WAP interface to unified messaging systems is the visual menu-based access to the user's messages. The user can access voice, fax, and e-mail messages directly from the unified mailbox with his or her WAP-capable mobile phone or communicator. The WAP access to a unified mailbox enables effective handling of large amounts of messages.

The access to voice and fax messages is presented in steps and illustrated from the user perspective in Figure 9.3.

1. The user is authenticated to gain access to his or her unified mailbox.

2. The message list is presented to the user containing voice, fax, and e-mail messages. The message list may include the deposit date and time, sender information, the size of fax files, the length of the voice files, subjects, and the possible attachments and size of e-mails.

3. The user selects a voice or a fax message from the list. After listening to the voice mail or selecting a fax message from the list, the user is presented with a menu of options for the message:

    ‣ Delete the message from the voice/fax mail system.

    ‣ Listen/print the message.

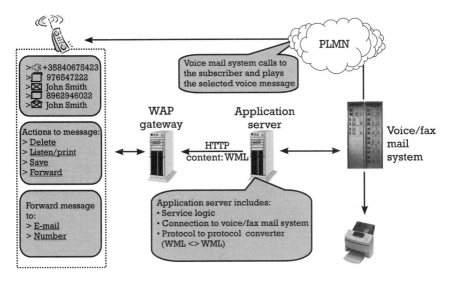

Figure 9.3   Voice and fax mail access.

> Save the message in the voice/fax mail system.

> Forward the message to other voice/fax mailboxes by giving the mailbox numbers, or e-mail by giving the e-mail address (possibly stored in the user's address book). The voice/fax messages may be attached to the forwarded e-mail messages.

From a network architectural point of view, the unified messaging system includes the following elements: a mobile terminal (with WAP browser), a WAP gateway, an application server, and a voice/fax mail server (see Figure 9.3.). The WAP gateway is the link between the WAP infrastructure and the Internet framework. It communicates with the WAP user agent on the mobile terminal using the WAP protocol stack over any of a number of possible bearers defined for WAP and with the application server using HTTP over TCP/IP. The application server includes a Web server for handling requests coming from the mobile terminal (passed through the WAP gateway which performs the communications protocol conversion). The application server also contains the user interface (the WML content and application logic) towards the mobile terminal user. The connection to voice/fax mail systems is often implemented using a proprietary protocol. WAP does not specify the interface between the application server and the GSM network elements (for example, voice/fax mail systems).

From the system perspective, the access to the messages in the unified mailbox is described with the same steps:

1. The WAP browser (on a mobile terminal) makes a request for the content of the message box (this request is forwarded through the WAP gateway to the application server).

2. The application server requests voice and fax messages from the voice/fax mail server.

3. The voice/fax mail server responds with list of messages.

4. The application server formats the response from the voice/fax mail server into a list (made with WML) and sends it to the WAP browser (as the response to the request made in step 1).

5. The WAP browser makes a further request including identification of the message selected by the user for listening (voice message) or printing (fax message).

6. The application server requests the voice/fax mail system to call and play the particular message to the requested phone number of the WAP user.

7. The voice/fax mail system initializes a call over the telephone network and plays the requested voice message (the data connection to the WAP browser is kept open simultaneously with the voice call).

8. After playing the message, the voice mail system terminates the call and the subscriber can continue with the message handling (delete, reply, forward, etc.).

9.6.2 Accessing e-mail messages

The most visible example of the explosive growth of the Internet is the wide acceptance of e-mail. E-mail has become the main messaging type and is part of the daily life of many people.

The worldwide installed base of e-mail mailboxes is now estimated to be nearly 400 million. The driving forces behind this growth are the increased Internet penetration and, more recently, free e-mail service providers. One of the key issues of e-mail access is that users want to be able to use their existing e-mail accounts (either corporate or personal) regardless of where they are (home, traveling, or in the office). Based on the large installation base of e-mail mailboxes, unified messaging systems must be able to support existing e-mail systems.

From the user's point of view, access to e-mail messages does not vary much from access to voice and fax message. The user must be able to (as in voice and fax messages) retrieve a list of (new) e-mail messages, read messages, and delete, reply, and forward messages. The architecture of an e-mail access service is also quite similar to that of voice/fax message access, but the different technologies and protocol used in the background to implement the service are quite different (see Figure 9.4).

From an e-mail server point of view, the application server acts as a normal e-mail client. The application server can use either IMAP4 or POP3 for retrieving messages from an e-mail server. In many cases IMAP4 is preferred, because POP3 does not have capabilities for message status handling. Using the POP3 protocol, it is, for example, impossible to keep the read/new status of a message up-to-date if the user is accessing messages from his or her normal e-mail client at work or at home while at the same time using the unified messaging services with a WAP phone.

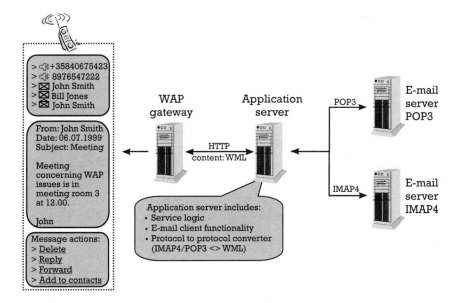

Figure 9.4　E-mail access architecture.

The technical flow behind e-mail access is as follows:

1. The WAP browser (on mobile terminal) makes a request for the content of the message box. The request is forwarded through the WAP gateway to the application server.

2. The application server makes an IMAP4 or POP3 request to the e-mail server(s). We are assuming the user has configured the e-mail server previously (for instance, through a Web interface).

3. The e-mail server responds with a list of requested messages.

4. The application server formats the response (from the e-mail server) into a list (made with WML, possibly also including voice and fax messages) and sends it to the WAP browser (as a response to the request made in point 1).

5. The WAP browser makes another request including the identification of the message selected by the user.

6. The application server requests more information of that certain message from the e-mail server.

7. The e-mail server responds with the message details (header, bodies, and attachments).

8. The application server formats the response (from the e-mail server) into a WML page and sends it to WAP browser (as a response to the request made in point 5). This response may also include possible actions for message handling (delete, reply, forward, etc.).

9.6.3   Directory services

Key supplementary services of unified messaging are directory services. The idea of directory services is to offer an electronic phone and address book that is always up-to-date. From the user point of view, directory service works as follows (see left side of Figure 9.5):

1. The user chooses to use a directory service. The directory service can be selected for example when the subscriber is composing an e-mail message or a short message, making a call, or forwarding a message to some recipient.

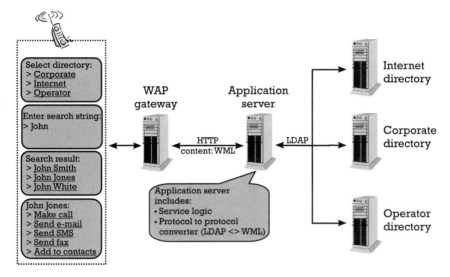

Figure 9.5 Directory access.

2. The user selects the directory he or she wants to use. Lists of directories can include public Internet directories, corporate directories (included, for example, in the corporate e-mail system), or operator directories (including, for example, the mobile subscribers of the user's home network).

3. The user also enters a search string. As a result, the user may get a list of entries in the directory matching that string.

4. The user selects the correct person. If the user was preparing to make a call when he or she selected to use directory services, that call can be initiated immediately after selecting the correct person (with the possibility to view the number before call setup).

5. The user selects the method for contacting that person (for example, "make call," "send short message," "send fax," or "send e-mail"), or he or she may proceed in order to get more accurate or detailed information about a person, or add that person to his or her personal contacts.

From an architectural point of view, WAP-enabled directory services include the following elements: a mobile terminal (with the WAP browser), a WAP gateway, an application server, and several directory servers (see Figure 9.5). The WAP gateway is the link between the WAP framework and the Internet or the LAN. It communicates with the mobile terminal using the WAP communication protocols and with the application server using HTTP over TCP/IP. The application server includes a Web server for handling requests coming from mobile terminals (through the WAP gateway), and a directory service client for making requests to directory servers. The application server also contains the user interface towards the WAP client user. The application server communicates with the directory server using LDAP over TCP/IP.

The technical flow behind directory service is as follows:

1. The WAP browser (on a mobile terminal) makes a request for a directory service. This request is forwarded through the WAP gateway to the application server.

2. The application server responds with WML content containing a list of possible directory servers for that particular user (all users are authenticated on the application server, for example, for billing and personalization purposes).

3. The WAP browser makes another request selecting the directory server and specifying the search string that is to be used to look for a certain entry (person).

4. The application server makes an LDAP request (including the search string) to the selected directory server.

5. The directory server responds with entries (containing personal contact information) matching the search string.

6. The application server formats this response from the directory server as a list (in WML) and sends it to the WAP browser (as response to the request made in point 3).

7. The WAP browser makes a third request including a specification of the person selected by the user.

8. The application server now responds with a list of actions possible for that particular entry. For example, if a telephone number was included in the directory entry, the application server offers the

possibility of initiating a call. There is also the possibility to add that entry to the user's contacts (located either in the mobile terminal or hosted on the application server).

When accessing a WAP phone's functions such as call control functions or address book handling, the directory service (or application server) uses parts of the WTA specification called WTAI (wireless telephony application interface). WTAI offers an interface for handling network functions (for example, call control) and WAP device-specific functions (for example, handling of an address book located on the mobile terminal). WTA and WTAI are discussed in more detail in Chapter 4.

9.6.4 Notification services

WAP offers a flexible way of notifying subscribers about new messages in their unified message box (including voice, fax, and e-mail messages). Notification services are important because the amount of messages that users are required to handle is rising all the time, and users usually want to be notified about important messages only.

In the WTA framework notification is performed using a so-called service indication (SI). SI provides the capability to send notifications to the WAP-enabled mobile terminal containing a short message and a link to a specific service. When the WAP browser of the user receives an SI, he or she can either start the service indicated by the link immediately, or postpone the SI for later handling. If the SI is postponed, the WAP WTA user agent stores it, and the user is given the possibility to act upon it at a later point in time.

From the user point of view, WAP notification works as follows (see the left side of Figure 9.6):

1. The user selects which messages he or she wants to be notified about. The user can select to be notified regarding only certain kinds of message types (voice, fax, or e-mail) and/or certain message attributes (e.g., sender, subject, message importance, etc.).

2. When a message satisfies the filtering rules set by the user and arrives in the unified mailbox, the user is notified with short text informing him or her of the event (possibly including the amount of new messages). This notification message may look, for example, as follows:

Your message box contains:
1 new voice message;
0 new fax messages;
14 new e-mail messages.

3. The user is given a choice either to either view the messages now (follow the link in the SI) or postpone the SI.

4. If the user selects to check the messages, he or she is forwarded to a WML page containing a message list in which the messages are sorted, for instance, by message arrival time (the messages that triggered the notification may be shown first).

5. The user can then read the messages that triggered the notification. Afterwards the user is able to read other messages as well and/or continue using other unified messaging services.

Notification services are often called push services. In contrast to pull services where the user has to actively request a response from the service, push services push information to the user when events occur that the user has indicated as being important in his or her profile or service configuration. The push framework introduces a new element, called the push proxy gateway, to the network architecture. The push proxy gateway forwards push requests coming from so-called push initiators to the WAP mobile client. Push initiators communicate with the push proxy gateway using the push access protocol (PAP). The push proxy gateway communicates with the WAP mobile terminal using OTA. WAP push services are described in great detail in Chapter 6. The WAP notification architecture is illustrated in Figure 9.6.

The technical flow behind notification service is as follows:

1. New messages coming from the voice/fax mail system or from the e-mail system (or at least an indication of the arrival of new messages) are forwarded to the application server. This forwarding can be done using SMTP, for instance.

2. The application server checks if this message is allowed to pass the filtering rules that are set by the user.

3. If the message passes the filtering stage, the application server sends an SI to the push proxy server using PAP. The SI includes

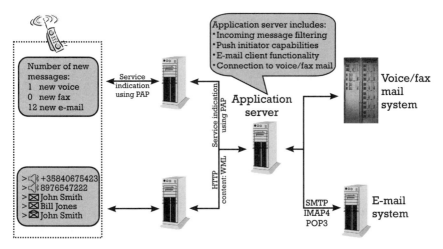

Figure 9.6 WAP notification architecture.

the number of new messages from every message type, so the application server must request the number of messages from other systems too. The SI may also include a direct link to the user's unified mailbox.

4. The push proxy server delivers the SI to the mobile WAP terminal using the OTA protocol.

5. When the user chooses to follow the link in the SI, a request for the message list is sent from the WAP terminal to the application server through the WAP gateway.

6. The application server responds with the message list including the user's messages from different messaging systems.

9.6.5 Service provisioning and billing

The network operator's customer-care expenses are increasing rapidly with the increased service offering and subscriber base. Service provisioning is one of the major functions that heavily increase the workload of the operator's customer-care operations. That is why service provisioning has lately received increased attention. With the huge success of the Web, self-provisioning has become a viable option to reduce the pressure on the customer-care operations. With WAP the self-provisioning becomes even more attractive when it is possible to manage the provisioning from the mobile phone by the users themselves.

Billing of unified messaging services is based on transactions; for example, every fax-mail retrieval or e-mail notification creates an entry in the billing log. The system keeps track of the end user's every action and collects the billing data. The collected billing data are transmitted to the operator's billing system for further processing.

The unified messaging service management and billing is implemented in the application server, which maintains a database for service operation and management purposes. This kind of architecture allows for the easy implementation of self-provisioning services. The application server is usually connected to a common operation and maintenance (O&M) center. The O&M center usually contains provisioning tools for the operator's customer care. Currently, Web-based provisioning tools are becoming a common way of implementing provisioning programs.

Billing data can be collected for every user action the user does in the application server. The operator can select the actions to be included in the billing and only those actions are then billed.

9.6.6 Self-provisioning with WAP

Advanced unified messaging services consist of a variety of services that can be taken into use separately. For example, e-mail notification or e-mail conversion to fax services can be offered to enable a more effective use of messages. Often the need for a new service comes up suddenly and the service must be put into use as soon as possible. For instance, the need for an e-mail conversion service could turn up before beginning a business trip without a laptop or access to e-mails. The fastest way to put the service into use is by self-provisioning of the service using a WAP-enabled mobile phone. The subscriber just accesses the self-provisioning tool using his or her WAP phone, selects the required service, and activates this service as described in Figure 9.7. After that the subscriber can use the service, and the initialization cost of the service is added to the user's phone bill.

9.7 Corporate unified messaging systems

As unified messaging services are mainly targeted to the corporate segment which normally already has a messaging system (such as e-mail or even unified messaging) offered by their companies, it is important to

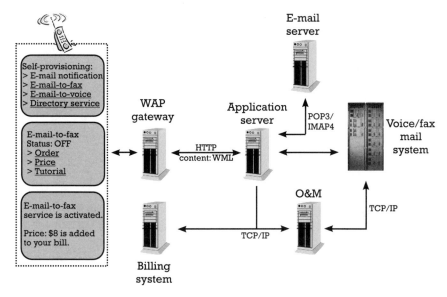

Figure 9.7 Self-provisioning with WAP.

also have access to these systems as well. It is characteristic for these corporate messaging systems to reside within the corporate intranets, which are protected by firewalls and are not easily accessible from the outside of the corporate LAN. IMAP4 or POP3 ports as well as HTTP ports for outgoing traffic are generally closed by the firewalls. This means that these systems cannot be accessed from the Internet in the normal way.

Another special characteristic for many corporate systems is that they may use proprietary protocols in parallel with the Internet standards such as IMAP4 or POP3 and that they can already offer a Web interface to the system. These systems generally also have many additional functions like calendar, address books and distribution lists, meeting reservation systems, task lists, and other corporate operative systems.

9.7.1 Network layout of the corporate unified messaging solution

Corporate solutions differ from a service or network provider's solution in the physical location of access servers and WAP gateways. Due to the fact that intranets often contain large amounts of sensitive data, they are generally well protected by firewalls and may also have additional protection using one-time passwords, access security measures, and other logging mechanisms. There are several different solutions for accessing the corporate data systems. Figure 9.8 presents a solution where circuit switched data (CSD) is used for connecting to the corporate remote access server (RAS) and a corporate WAP server is used for making the delivery of WAP content. Still, an application server is needed for accessing different e-mail, LDAP, and other servers, if they do not have a WAP interface in place already.

The modem connection for dialing in can be provided by either the operator or an ISP, but from then onwards the link must be secured by using virtual private network technologies, for instance.

The short message service can also be used as a bearer for accessing the corporate information systems. However, in that case the network layout could be different. In Figure 9.9 the corporate intranet is connected to the operator's short message service center (SMSC) in order to use SMS as a WAP-content bearer. The SMSC must support a feature where all short messages which are sent to a service number are diverted to a specific IP address. Note that this function may also be performed by dedicated gateways. The SMSC must also support TCP/IP connections to the third-party service providers as well as allowing for secure traffic between parties.

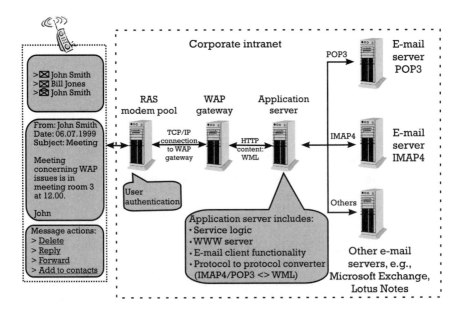

Figure 9.8 Schematic representation of the connection between the WAP gateway and the corporate remote access server using circuit switched data.

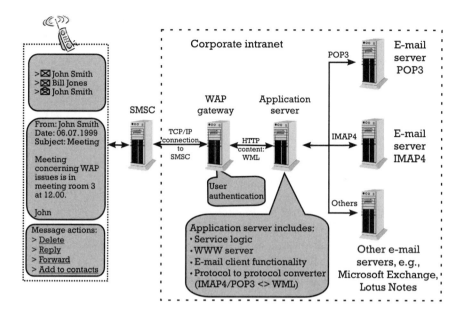

Figure 9.9 Connection between SMSC and corporate networks.

9.7.2 Wireless application protocol messaging and additional services in corporate systems

Despite the fact that corporate networks represent a different situation compared to operator systems, they can offer pretty much the same functionality to end users. Most of these services are already described in the previous sections. Many corporate e-mails systems, however, contain some additional features that are summarized here:

▸ *Calendar.* Access to the corporate and private calendar in the corporate server. With this service users always have only one calendar to access. They do not need to have separate calendars in desktops and the mobile devices they use when on the move.

▸ *Public folders.* Public folders are features in e-mail systems where all people have access to the same e-mails stored in the folders in an e-mail system. Access to such folders can also be restricted to a closed user-group, for instance.

- *Meeting agent.* Meeting agents help workers in big companies to make meeting appointments (as well to reserve meeting rooms) and send messages to the people who should be attending that meeting.

- *Task lists.* Personal task lists are lists containing important tasks to be done with their descriptions and time schedules. They are typically included in calendar systems.

- *Phonebooks, address books, and contact lists.* With WAP services users are also able to access their address books, contact lists, and other directory services provided by their corporate networks.

- *Operative systems like enterprise resource planning (ERP) applications.* Corporate users can also in principle be able to access ERP systems through a WAP interface. Services provided by ERP systems can be, for example, checking the status of delivery or stock level of an item as well as price lists and project time schedules. Given the rich nature of the information flow within a large company, the number of possibilities can appear endless.

All the above-mentioned services can in principle be accessed via corporate-specific WAP gateways in a secure way. These systems must have a front end providing an interface to the Internet through a Web server, which then can be connected to the corporate WAP gateway to provide WAP access to such corporate information systems.

Contents

Mobile Financial Services and Applications

Stuart Marsden

10.1 Introduction

This chapter looks at the opportunities and challenges facing the designer of mobile financial services and applications. We will explore the potential services that may be offered and the issues that surround their implementation.

While this book is about WAP, in this chapter WAP is put into context by exploring the alternatives. While the technical details of WAP are covered in the rest of this book, this chapter is seen through the eyes of an application developer.

10.2 A new electronic channel is born

It is only very occasionally that an application developer has the opportunity to contribute to the early pioneering stages of a completely new computing paradigm. The fact that you are reading this indicates that you are at least considering it. For the past 25 years, data have been confined to the province of first metals and then semiconductors. This is all about to change. Early "clunky" and slow to initialize GSM modems are now about to be replaced with a complete family of compact wireless devices. Probably the most prevalent of these will be the WAP-enabled phone.

The current rapid growth of mobile phones and the phenomenon of the Web are about to collide and release enormous amounts of potential energy. The finite spectrum for carrying data wirelessly has already become a scarce and hence valuable resource. One may even find that someday 1800 MHz futures are traded along with other commodities.

However, every silver lining has a cloud. In the emerging world of next-generation mobile applications, it will be the power struggle between network providers and handset manufacturers to control the channel. Content providers, Internet portals, and financial institutions will maneuver for the best position to exploit the channel.

Technology suppliers, eager to gain a foothold in this brave new world, are hyping their current tools and projects more than usual. There is also a built-in human tendency, when looking at exponential growth in a technology, to overestimate the initial growth rate, but equally to underestimate the final uptake.

Even allowing for the inevitable nature of the evolution of the channel, the growth of this market will be spectacular. Gartner Group is now predicting 600 million phones by 2001 [1]. This has attracted the big players into this market, most of the leading system integrators and large consultant companies, and all of the leading financial services organizations already have a strategy to exploit mobile commerce. At the time of writing, Steve Balmer, CEO of Microsoft, announced that it was joining the WAP Forum [2], as Microsoft transitions from a PC to a device-centric world. Articles about wireless financial services are starting to appear in the press every week (e.g., Virgin, with its strong brand, has formed an association with the United Kingdom wireless network operator One 2 One).

10.3   Who are the users of this new channel?

Eventually it will be everybody, especially given the spectacular success of "pay as you use" schemes in the United Kingdom (uptake figures not generally published) where subscribers regularly top up their phone balances using mobile commerce techniques. In due course, familiar and trusted retail brands may become the norm as branded wireless network providers. In the meantime, mobile financial services and applications will target the high net-worth individual and the mobile executive. Figure 10.1 shows the probable distribution of mobile workers, excluding teleworkers who are permanently located at a single off-site location. This indicates that 25% of office workers (nomads and mobile managers) spend a significant portion of their time traveling. In addition to this, we have the business traveler; research shows that more time is being spent traveling, increasingly on pan-European trips. Finally, there is the commuter; and again the trend is towards an increase in the time spent commuting.

The overall trend is for mobility to increase. This has the effect of reducing the amount of leisure time available to target users. Hence, any tasks that would normally eat into leisure time that can be completed while on the move will be a good candidate for early adoption. In these circumstances it is convenience and general availability, rather than cost, that will dictate the uptake of any new service.

10.4 Previous constraints to mobile commerce

As with any new channel, there have already been some false dawns, the principal inhibiting factor having most commonly been security. Analog mobile phones are not considered to be secure by the banks, such that the small print of many telephone-banking services prevents their use. The whole area of authentication and the lack of legally recognized

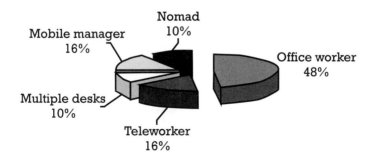

Figure 10.1 Mobility of workforce (source: Matt Schofield).

personal digital certificates have forced the GSM SIM (subscriber identity module) card to become the de facto standard. The problem is further compounded by the establishment's desire to be able to eavesdrop on transactions using copies of private security keys held in escrow by some trusted authority for the good of the nation.

The adoption of smart cards for the storage of e-cash (used here as a generic term for electronic cash technologies or systems) has also had a slower-than-expected uptake. Trials that have taken place to date have been less than spectacularly successful.

Finally, many financial establishments are suffering from what can only be described as "channel fatigue." After the traditional branch-and-call center channels, we have now had, in a relatively short time, multi-media kiosks, the Internet, personal financial managers (Quicken and Microsoft Money), interactive TV, and now wireless, that have all been competing with European currency introduction and year 2000 projects. Unless an integrated multichannel architecture is adopted, every new channel increases the time to market for the launch of new products.

10.5 Breakthrough technology

According to the ARC GROUP, the number of mobile subscribers in 1999 (428 million) greatly exceeded that of Internet users (241 million), and is expected to grow to more than 1 billion by 2003. The general availability of browser-based mobile phones will mean that mobile browsers are likely to become the most prevalent devices on the Internet. An analogy is that SMS was the equivalent of FTP in the Internet, and it will be WAP, the equivalent of the Web, that will cause the explosion in mobile applications.

In the short term it will be integrated message services, where the Internet, voice, PDA, fax, and wireless technologies can all be seamlessly integrated, that will provide the "very menacing," if not "killer," application.

10.6 Strengths and weaknesses of the mobile channel

The strengths of the mobile channel centers on extending the mobile phone are the basis of new devices. First, the mobile phone is becoming ubiquitous, as indeed, in some countries it already is. The compact size of modern mobile phones is such that they are now an "unconscious carry," unlike laptops and most PDAs. This will promote their widespread use and extend the time they are switched on and in close proximity to their owners. They will in effect become an essential digital extension of their owner. This makes the phone the ideal platform for value-added services that provide timely information.

It is a virtually "instant on" device, the biggest delay being entering SIM number and/or phone-based PIN. It is possible to get a bank balance by SMS message while a PC is still booting, never mind the additional time taken getting into on-line banking. Using preprogrammed keys, it is possible to have "one click" services delivered to the phone, such as traffic or financial information. The GSM digital network is now secure enough for all but the most security-conscious applications and organizations. The mobile phone display and, hence, SMS messages are inherently private. They can be delivered to an individual and can be discreetly read in most circumstances. When location information is made available to the application by the network, it is possible to offer a level of personalization not available with other channels.

The phone smart card has the potential to store customer and other data, which can be accessed by the wireless application. The next big breakthrough (probably during 2001) will be when mobile phones generally have a second smart card. This, together with resurgence in the uptake of e-cash, has the potential to give everybody an ATM in the pocket.

The weaknesses of the channel arise from the device and the network. There are ergonomic considerations: for example, it is not possible, without the use of a hands-free device, to use the voice channel at the same time as either the keypad or the display.

Many people perceive the network costs as high, and the precedent has been set that value-added services are chargeable on the mobile, whereas they may be "free" on the Internet. Care must be taken when developing mobile applications not to abuse the use of underlying chargeable services such as SMS, without the user being aware of the costs involved. The smaller bandwidth and higher latency of this medium compared with the Internet mean that content has to be optimized. It is also desirable for the mobile application to use asynchronous messages where possible, if we are to maximize the perceived responsiveness of the system, in line with good usability practice.

The "mobile user on a train entering a tunnel" phenomenon, apart from inciting wireless rage from the user and bringing smiles to the faces of the rest of the passengers, is a good example of the inherent unreliability of the connection. In addition, there are still many locations where the service is still patchy. Hence, good crash recovery with transactional integrity is essential for any serious mobile financial services application.

The limited display means that personalization is essential to eliminate unnecessary options and data from the WAP application. The keypad is a device designed for only limited input, and therefore until voice recognition or predictive typing becomes common, it is essential that the user is not asked to key any information that could reasonably be known or deduced by the financial services organization.

There is now a conflict of requirements for the handset. As a phone, the trend is to make the handset smaller, cheaper, and have a longer service life between charging. As a data terminal, the trend is to have bigger displays, faster local processors, higher bandwidth, and more data messages, all of which will reduce battery life and drive up handset costs.

The net result of this is that we will see an increasing variety of wireless data devices. It will become increasingly important that a standard, such as WAP, shields the mobile application developer, as much as possible, from the vagaries of the individual devices.

10.7   The current range of mobile devices

10.7.1   SMS messages on GSM phones

This is currently the most prevalent way for mobile financial applications to communicate with the handset. It is relatively simple for an application server on the network to generate SMS messages, which can prompt for a reply. The message body of the reply must be entered using the keypad, while the return address (e.g., the telephone number) is generated automatically. It is feasible to construct a simple reply mechanism, such as:

▸ Messages containing the number one mean a request for more information.

▸ Messages containing the number two would indicate confirmation, etc.

SMS messages are limited to 160 characters; however, in practice, given the limitations of the keypad and screen capabilities, this should be enough for most applications. It is also possible to achieve a similar effect using an SMS/e-mail gateway. However, in this case some of the message is given over to e-mail fields such as subject, and automatic reply is not as easy, as the e-mail address must be entered in the reply. Figure 10.2. is an example of a simple e-mail message sent via SMS.

These services can be general in nature, such as news broadcasts: localized, possibly using cell broadcast techniques, or personalized such as registered alerts for share-price movements. It is also possible to use pre-created messages, which trigger an automatic voice message reply, such as traffic updates.

10.7.2 USSD messages on GSM phones

Unstructured supplementary services data (USSD) GSM phase 2 messages are similar to SMS-based messages; however, they work in a different way. A connection remains open for the duration of a session, which removes the need to reserve a channel for each message as is the case for SMS. The result of this is that USSD messages are many times faster than SMS and have a reduced latency. USSD messages are used to activate supplementary services such as call diversion using strings such as: *#21*nnnnnnn#SEND, where nnnnnnn is the number to divert to.

At present only the network operator can process USSD messages. Current SIM toolkit applications are not designed for USSD; therefore, it is not likely to be used as general message service. USSD, however, will become a bearer of WAP, despite the fact that USSD is half-duplex, and it is therefore important that the application developer is aware of its existence and limitations.

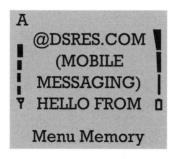

Figure 10.2 Simple e-mail message sent via name@airmail.co.uk.

10.7.3 Applications on the SIM of the phone

Known in the industry as SIM toolkit or STK applications, this option requires GSM phase 2+ enabled phones, which are now becoming more common and likely to become very common if STK-based applications become widespread.

The applications are written using the development environment provided by the SIM manufacturer. The most commonly used STK development tool is manufactured by GEMPLUS, which has a GUI tool for building application menus, but still requires serious development effort to develop the application logic. The application (called a masked function) must be installed on the smart card, along with the SIM components from the network provider. Hence, this option is really only feasible with the active cooperation of the network provider. The application is therefore effectively tied to the SIM manufacturer and the network provider. It is now possible to download updates to the application (called filtered functions) from a server using a technique known as OTA, or "over the air." However, this server must reside on the network and thus is under the control of the network provider.

10.7.4 Microbrowser in smart phones and PDAs

There appears to be three main types of microbrowser available.

1. *Wireless markup language-based browsers.* Browsers that can render content formatted using WML will be present in the WAP phones, the primary subject of this book, and referenced here for completeness. Chapter 2 discusses in detail WAE, of which WML is an element.

2. *HTML-based browsers.* These devices provide a smaller version of a full-blown Internet browser. For instance, the United Kingdom company STNC (http://www.stnc.com) has developed a browser which is embedded in larger communicator phones such as the Philips Accent, and PDAs (personal digital assistants) such as the Psion 5. Figure 10.3 shows the browser in a large phone form factor. Some vendors (e.g., Compaq) have also developed dual-mode WML/HTML browsers.

    The Windows CE devices come with Pocket Internet Explorer. These classes of browsers will always offer reduced capabilities compared with Internet browsers. The widespread adoption of these browsers and their success will depend on two factors. The

Figure 10.3 STNC browser (picture courtesy of STNC).

first is the perceived target platform for popular Web sites. If this was to move away from frameless browser support and incorporate advanced features such as DHTML, Java, and Shockwave, mobile HTML browsers will be left behind. The second factor is the quality of the "translation" of the current Web sites to match the limitations of mobile phones. Clunky performance or the need for excessive screen scrolling will detract from their use, especially on a device that also has limited input support.

3. *"Web clipping" based Palm VII.* The Palm VII, at the time of writing, is in trials in the United States. The approach that 3Com has taken is a hybrid solution between generic Web browser and the wireless targeted WAP. Rather than general Web browsing, individual forms representing mini-applications (Web clippings) are preloaded on the Palm VII. The forms generate standard HTML queries, and the Web site response is converted by a Web-clipping proxy server into a form suitable for transmitting to the Palm VII. This approach is similar to that of HTML and WAP gateways (see Chapter 5 for technical details on WAP gateways), with the exceptions that the Web clippings are preloaded and the technology is obviously proprietary to 3Com.

10.8   Resident applications on the mobile device

In this case the application resides on the device as opposed to on a smart card, as in the case of the SIM toolkit. This category also extends to PDA and phone combinations. For the smaller devices the EPOC operating system has become popular (for instance, the Ericsson R380 phone), while Windows CE is used in many of the palm-sized and handheld computers that do not yet have phone functionality.

In these cases the application is downloaded from a PC and installed into nonvolatile memory in the device. The primary data access mechanism is synchronization with a personal information manager (PIM) such as Microsoft Outlook. Applications are written using standard, or slightly modified, development tools such as Visual Basic for Windows CE. Where the PDA does not have built-in wireless capabilities, an IrDA (infrared data association) link to a phone is becoming common. However, this requires that the two devices are suitably "arranged" for use, which is not always practical when traveling. This problem will be solved with the arrival of Bluetooth [3], a short-range wireless personal LAN for device interoperability. The functionality is rich, but not very portable. The extreme of this class of devices is the laptop running Windows NT or Windows 98 with either a GSM modem or IrDA link to a mobile phone.

10.9   Choice of mobile commerce platform

The choice of a mobile commerce platform will depend on many factors, including time to market, and whether the application has to be "broad reach" or can be constrained to run on a specified range of devices. Figure 10.4 shows a comparison of the platforms.

From this analysis it appears that, assuming that the content is created or transformed from the Internet and politics do not get in the way, WAP is likely to become the dominant mobile platform.

10.10   Existing mobile financial services and applications

Mobile financial services and applications are being launched with increased frequency using a variety of technologies. Several applications are voice based, using DTMF or speech recognition to interact with an

	SMS messages	SIM toolkit based application	PDA and phone combination	HTML browser in phone	WML browser in phone
In general circulation now	Yes	Yes	Yes	No	No
Broad reach by 2002	Yes	On 2nd SIM card	Yes	Yes	Yes
Open standard	Yes	No	No	Yes	Yes
Available across a range of devices	Yes	No	No	Yes	Yes
Over the air downloadable	N/A	Only by network operator	No	Yes	Yes
Optimized for wireless	Yes	Yes	No	No	Yes
General availability of content	Some	No	Applications rather than content	Yes	Predicted to be yes

Figure 10.4 Characteristics of mobile platforms.

interactive voice response (IVR) system. The IVR can play prerecorded information with embedded dynamic data. In addition, the IVR can also generate SMS or USSD-based messages, where a message in a text format is required.

An example of this is the Co-operative Bank in the United Kingdom, which is getting more than a 1,000 calls a month for mini-statements. Co-op research shows that these requests are typically made to check balances before a significant purchase. The calls are subsidized and are generated by a call to a short number of the wireless network operator Vodafone. Figure 10.5 shows the text of the resultant SMS.

The Co-op mini-statement is an example of pull technology. An example of push technology is the FT Cityline service operated by the Financial Times in the United Kingdom (see Chapter 6 for a more technical discussion of push related to WAP, and Chapter 9 where the use of push services in unified messaging systems is outlined). In this case the caller gives the IVR the code for the required stock and a trigger level. When there is movement in the stock large enough to hit the trigger, an SMS message is sent to the phone. Figure 10.6 is an example of such a message.

18May1999 CURR1234-Bal £741.09CR, Last 4Items, £29.47DR, £91.68DR, £16.16DR, £2.00DR, -O/D Limit £NIL.

Figure 10.5   SMS-based mini-statement.

FT cityline alert *** AIT group *** n/c at 365, bid=362, ask=368, vol=0, at 9:44 *** "0374519971" ***

Figure 10.6   SMS message from FT Cityline.

Products such as X.SMS Banking from Brokat allow the SMS messages to be digitally signed to authenticate the sender. This is implemented using an application on the SIM. It can be used to secure messages sent via wireless. Alternatively, high-value transactions to a call center, for example, via a landline, can generate a secure SMS message to which the user can respond after entering his PIN on the phone. In this way the user is effectively confirming the transaction via a digital signature (combination of possession of the SIM and knowledge of the PIN).

In 1997 the United Kingdom wireless operator Cellnet and Barclaycard launched a SIM-based application that allowed the easy generation of balances and mini-statements via SMS. The application was launched using a dedicated "B" button on special phones branded by Barclaycard. It was reported that the service had 120,000 subscribers after one year.

Also in 1997 the Swedish bank Postgirot and Swedish operator Telia Mobitel launched an application called MobilSmart® based on a GEMPLUS SIM which allowed for banking transactions and utility bill payment using just a mobile phone. In August 1998, Expandia Banka in the Czech Republic launched a dual SMS (using keywords) and SIM toolkit-based banking application. The screens for a balance request are shown in Figure 10.7.

In January 1999, Citibank Singapore and the local mobile operator M1 also launched a mobile banking service. The SIM toolkit application downloads personalized menus based on the services and accounts the user holds with the bank. In addition to balances and mini-statements, most of the features found in Internet banking are available via the phone menus. The messages are encrypted with a key length of 128 bits using DES encryption.

(a) (b)

(c) (d)

(e) (f)

Figure 10.7 Balance inquiry using SIM toolkit from Expandia Banka; (a) in the main menu select menu item Expandia, which will highlight and confirm it; (b) in the next menu, displayed on the phone, choose item Info (information about account) and confirm it; (c) the next menu will be displayed, from which you can choose required service (balance). Selected item will highlight, confirm it; (d) then you will be asked for your account number. When you have entered the number, confirm it by OK; (e) display will show animation about sending of message with your request; (f) when the transmission is successful, this is confirmed. After a while, the phone will receive the requested information.

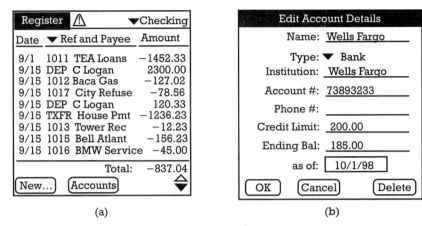

Register ⚠ ▼Checking	
Date ▼ Ref and Payee	Amount
9/1 1011 TEA Loans	−1452.33
9/15 DEP C Logan	2300.00
9/15 1012 Baca Gas	−127.02
9/15 1017 City Refuse	−78.56
9/15 DEP C Logan	120.33
9/15 TXFR House Pmt	−1236.23
9/15 1013 Tower Rec	−12.23
9/15 1015 Bell Atlant	−156.23
9/15 1016 BMW Service	−45.00
Total:	−837.04

[New...] [Accounts]   △▽

Edit Account Details	
Name:	Wells Fargo
Type:	▼ Bank
Institution:	Wells Fargo
Account #:	73893233
Phone #:	
Credit Limit:	200.00
Ending Bal:	185.00
as of:	10/1/98

[OK] [Cancel]   [Delete]

(a)   (b)

Figure 10.8 Sample screens from Pocket Quicken from Landware running on the Palm Pilot; (a) shows a list of transactions for the selected account, and (b) shows how to create or edit a transaction.

The functionality of mobile financial applications increases rapidly as we move on to the PDA. Figure 10.8 shows some screens from Pocket Quicken from Landware, running on the Palm Pilot. This application works off-line and is then synchronized with a PC. It is likely that this screen resolution will eventually be available on WAP devices.

There is already a gradual trend for off-line mobile financial applications (such as Money and Quicken) to become more Web based. This, together with the conclusion in the last section that WAP is likely to be the dominant wireless technology, suggests that over time all personal finance mobile applications will migrate to use server and WAP-based technologies.

10.11   Principles of building scaleable *n*-tier applications

Building industrial-strength, mission-critical applications that perform well, are resilient, and scale to a large number of users will challenge the most experienced development team. However, with more than 10 years of experience of building distributed and *n*-tier applications, it is possible to distill the most important principles into a short list of recommendations that are applicable to most application developments. It is important

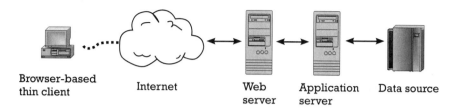

Figure 10.9 Typical *n*-tier application topology.

from the outset to consider not just the development but the whole life cycle of the application. Figure 10.9 shows a typical thin-client configuration.

The first principle is that the server should be component based. There are several demonstrable benefits (manageability, malleability, reuse, etc.) of using components in traditional client/server applications. These still apply in *n*-tier. In addition, most commercial application servers define some form of bundling code to ease administration and deployment of the application. Collections of components are a great way of defining these bundles. It is important to get the granularity of the components correct. As a rule of thumb, small applications have less than five components, while very sophisticated applications may have as many as 20 to 30 components.

It is possible that different components within a single application may be written in different languages—both Microsoft Transaction Server (MTS) and CORBA (Common Object Request Broker Architecture) support this, for example. The component boundaries and interfaces will be defined as part of the design process. The components normally represent a significant body of functionality that can be treated as a black box. Examples of this would be a work flow or a customer contact subsystem. However, it is also good practice to create small components which encapsulate functionality that may be subject to frequent change. In this case, the goal of defining a component is not to reuse but to insulate the rest of the application from the impact of that change.

The second principle is that while the whole application should be sourced from the server, functionality must run in the most appropriate location (client or server). The biggest challenge to thin-client developments is to avoid an overreaction to the problems of traditional fat clients. This results in very thin or "anorexic" clients. The prognosis for these applications is not good and the symptoms are listed here.

‣ Slow—Excessive interaction with the server will significantly degrade the performance and hence usability of the application.

‣ Excessive network traffic—This is even more acute over a slow network with high latency, or where network charges are based on packets transmitted.

‣ Poor scalability—The server has to participate in every interaction with the user and hence can support fewer clients compared with a more distributed application.

The ability to deploy client-side application logic will depend on the particular client platform. Where client-side scripting is available, this should be used to perform simple functions such as field-level validation. For mobile applications this will become more important as dual-slot phones start to ship with a second SIM being able to run a variety of applications.

The third principle is that the client must be self-installing and maintaining in order to deliver a key benefit of n-tier: lower cost of ownership for the clients. Therefore, attention must be paid to versioning and deployment of the application. This would obviously be critical in a SIM toolkit-based application.

The fourth principle is that an efficient mechanism is required to pass data between the physical locations of the software (i.e., between client and server and between multiple servers). The extensible markup language (XML) [4] is becoming a standard in this area; however, care must be taken if the XML is not to become very verbose.

The fifth principle is that proper error handling and logging is vital. As there is a network involved, it cannot be assumed that any operation, no matter how simple, will succeed on a remote device. It is essential for a financial application that the state of the application or data are never ambiguous and that the user is always aware of the success or failure of any transaction. This brings the developer into the world of two-phase commits and guaranteed notification.

The sixth principle, related to the previous one, is that the applications developed must be capable of being debugged and maintained. Therefore, tool selection and the ability to debug the complete application in a near-live multiuser environment is vital.

The seventh principle is that the server architecture, which is going to be fatter than traditional client servers and potentially supporting thousands of clients, must be very scaleable. There is a lot of work being done in this area by the application server developers to assist the application developer. However, it is impossible to shield the world from poorly designed or implemented applications. The techniques involved in this area will depend on the underlying technology (e.g., MTS, CORBA, or Enterprise Java Beans (EJB)).

The eighth principle is that security must be designed into the architecture to allow applications to run across the relevant public network.

10.12   Building WAP applications

With the introduction of WAP, the development issues are compounded by the immaturity of the tools and lack of extensive experience of the mobile phone as a data channel. Figure 10.10 shows a typical WAP server configuration with *n*-tier application topology.

The good news is that WAP development will be very familiar to Internet developers. The concepts of URLs, application servers, markup languages, and scripting languages carry across to WAP. WML and WMLScript are covered in Chapter 2, and hence they will not be discussed here. Instead, we will look at some of the issues facing the WAP application developer.

Figure 10.11 shows the logical layers in a typical WAP client. This is similar to that of a PC-based browser, except that the WAP-based application will be able to exploit the SIM's security and storage facilities in the future releases of WAP.

The first obvious aspect of WAP development is that it is currently very difficult to build and deploy a real end-to-end system. At the time of writing, WAP-enabled phones and network access are very limited. This will become much easier over time as we move from "bleeding edge" to "leading edge" and then mainstream development. In the meantime, the average developer must be content to work against WAP emulators

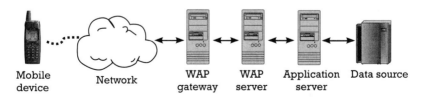

Figure 10.10   Typical *n*-tier WAP application topology.

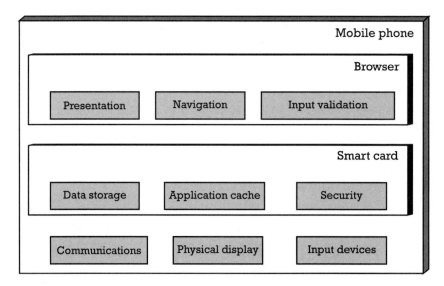

Figure 10.11 A WAP client structure.

running on PC or UNIX clients. There are several WAP toolkits available, with varying degrees of completeness and conformance to the WAP standards.

While vanilla WML is covered well by these toolkits and servers, topics such as session handling and application state management are not as well addressed. It is in these areas that the Internet developer will start to find himself or herself in unfamiliar territory. Browser cookies, an integral part of Internet development, have been replaced with long-lived browser variables that survive changing WML card decks. It is necessary to refer to the WAP specification (see www.wapforum.org) to find more information.

A big design issue for a WAP developer is whether to build new WML content or try to parse and scrape existing HTML content. The author is convinced that the latter approach is not sustainable over the long term. Instead, it is possible that content providers will embed WML content in Web sites. The delivery of a specific content format will be conditional on the browser type just as frame or ActiveX support is now. However, while this approach has the advantage that all the content is colocated, it is in effect just partially duplicating the content and increasing the amount of regression testing required when a new page is produced. A more sophisticated approach, and definitely more effort initially, is to derive both the

WML and HTML from some common device neutral markup language. This approach has the advantage that the presentation for any particular content is only defined once and then just automatically rendered using the appropriate technology. These approaches to WAP application development are outside the scope of this chapter, so for the sake of this discussion, we will assume that we are just creating vanilla WML.

The next issue is that WML is still evolving and therefore version issues (e.g., 1.0, 1.1, 1.2, etc.) of the standards will need to be treated with care. In the short term this may not be an issue if version 2.0 becomes established before any significant number of 1.1-based phones are shipped.

WAP is playing an excellent role in standardizing the mobile device platform. However, it is likely that, just like the Internet, vendors will "embrace and extend" the standard in an attempt to differentiate themselves and "add more value" to the platforms. Unfortunately, unlike the Internet, new browsers cannot be easily downloaded and upgraded. The network operators are also unlikely to subsidize the rapid change of handsets, which are already heavily subsidized for subscription-based mobile customers. This may well result in an increased diversity of handset browsers in the market and the issues of support for legacy WAP browsers. This will inevitably lead to a de facto standard for the lowest common denominator WML, with browser-specific extensions conditionally generated by the WML server on the basis of user preferences and/or browser capabilities.

It is well recognized that the WAP device has limitations compared with a PC-based browser. However, apart from the obvious limitations of screen resolution and network bandwidth, other limitations may also apply. For example, the dynamic memory available will be restricted, and this will in turn limit the amount of WML that may be downloaded. Depending on how the memory is allocated, it may be possible to download a WML page that loads correctly, but then fails if the WMLScript allocates too much memory. It is not clear how any particular device will behave in these circumstances or how these errors can be detected or trapped by the application.

As with the PC browser, it may be possible to deduce some characteristics of the WAP device from the HTTP header information. However, as the range of WAP devices increases, this is unlikely to contain all the information required to decide on the content and formatting of individual WML pages. One approach would be to ask for the phone model when the user first registers for the service. However, it should always be made possible to amend these settings later, especially as it is easy for the user to swap phones, while continuing to use the same SIM with the network left unaware of the change. Work is underway within the WAP Forum to allow the transfer of user preferences and browser capabilities but such a standard, let alone implementation in the actual WAP phones, is not yet finalized.

10.13   Building multichannel applications

In today's financial services environment, it is unlikely that any mobile commerce application will be built in isolation. In fact, it is now becoming normal to consider new devices such as mobile phones or interactive TV as part of a larger (emerging) channel strategy. These projects are typically initiated by marketing or business units within a financial services organization, often with the help of external channel specialists. All too often a successful prototype will lead to rapid development and deployment. Unless there is already a multichannel infrastructure in place, this approach to development will lead to a stovepipe implementation, as can be seen in Figure 10.12.

This type of solution often leads to several problems. First, the costs will be higher taken over the development of several channels compared with an integrated approach, as a significant amount of functionality is duplicated. Second, the channels may have a different look and feel, and at one extreme will have inconsistent branding and financial product representations. The third and most significant issue is that it becomes harder to launch new products, as they must be supported across all the disparate channel implementations. To alleviate these problems, it is now becoming popular to introduce common business objects across all channels. The channel-specific presentation layers are then created using the appropriate technology for that channel. This is shown in Figure 10.13. More information about the separation of content and formatting can be found in Chapter 2.

This approach can be extremely effective, but there is a potential danger on the horizon, which will cause even this approach to have problems. We are now seeing increasing fragmentation and diversity within the channels. Not very long ago, we were talking about interactive TV (iTV) as one channel. In reality there are now satellite, cable, terrestrial, and asynchronous digital subscriber line (ADSL) phone lines. Each of these subchannels may have their own technology for implementation,

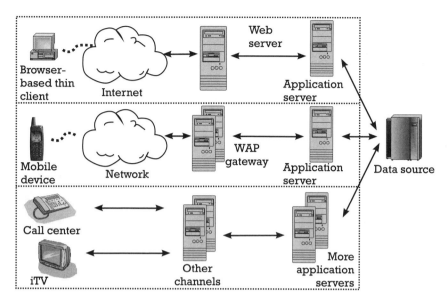

Figure 10.12 Typical stovepipe solution to multichannel support.

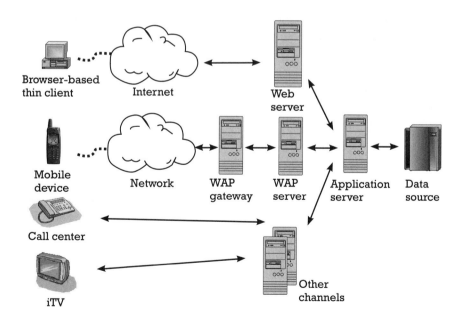

Figure 10.13 A shared business object: multichannel delivery architecture.

which may or may not be proprietary. Equally important, each of the subchannels has different operational characteristics (e.g., the back channel of iTV may be permanently open or not), which needs to be taken into consideration when designing applications.

A similar situation exists with the mobile channel. As we have already seen, we are faced with the prospect of dealing with SMS and USSD messages, voice or DTMF-based IVRs, SIM toolkit, WAP, PDAs, and laptops. While there is a good chance that WAP will become the dominant technology for mobile devices, it will be several years before it is ubiquitous in the installed base of mobile phones. In the meantime, we are faced with a heterogeneous and fairly diverse environment. It is therefore probable that the mobile commerce provider will have to be content with at least two of the above technologies.

In this world, with many subchannels to support, it is desirable to eliminate as much as possible the business functionality from the channel-specific parts of system. In the current model, the generation of the presentation layer from the underlying business objects is done on a channel-by-channel basis. While this code is not extensive in a thin-client environment, there is nevertheless duplication and replication of logic. For larger, new developments it will probably be worthwhile to introduce another tier in the architecture. Figure 10.14 shows a new channel-neutral presentation layer.

In this case, rather than trying to postprocess the Web interface into other formats, all the channels derive their presentation layers from a new channel neutral format (CNF). In this model, a single logical process constructs the CNF for all channels. This is then parsed and rendered for each (sub)channel. In this way the channel-specific code becomes more stable and does not need to be changed when the application is enhanced or new application features are added. An increasingly popular approach to this is to use XML for the CNF. This XML-based intermediary presentation definition has several advantages.

- Web browsers are evolving to be able to render XML directly by using cascading style sheets (CSS) and extensible stylesheet language (XSL).

- WML is available as an XML document type definition (DTD).

- Parsers and other XML-based tools are now readily available in several languages (Java, C++, and Visual Basic).

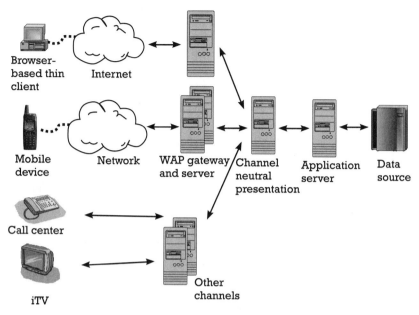

Figure 10.14 A next generation multichannel delivery platform.

▸ XML is becoming the language of e-commerce, with existing standards such as OFX migrating to be XML compliant.

▸ XML is easily extensible, so addition tags can be added such as MIN|MAX values or mandatory indicators on requested data fields. In this case care must be taken when parsing the XML, so that unexpected new tags are properly handled.

10.14   Building financial WAP applications

Over and above the considerations for building WAP applications, there are several factors which must be considered when designing and constructing mobile financial services (FS) applications.

It is likely that early FS applications will target mobile business people (as described in the introduction to this chapter) and high net-worth customers. These are exactly the same people who will adopt other value-added services targeted at them. It is important therefore that any new FS application should at least coexist with, and preferably cooperate with, these other services. The most important of these is unified messaging.

With unified messaging it is possible to mix and match message creation, notification, and delivery to the most appropriate device at the time. In this new world users can be notified on a mobile phone of the arrival of e-mail in their universal or corporate inbox from a VIP sender. They can then opt to be sent a synopsis, forward it to a colleague, or send a copy of the text to a local fax machine. They can even elect to have their e-mail read to them, and choose to dictate a reply into an IVR-enabled message center.

The wireless network providers are embracing this goal by designing "intelligent networks." With a single number for all of my calls, land and wireless, the network can be told when I am working, traveling on business, or enjoying leisure time, and respond accordingly.

Sophisticated mobile users will soon expect that all mobile commerce applications will work in a compatible way to their most used value-added service: messaging. As an example, I should be able to buy shares using my WAP phone, hold the contract note as a message, and print it once I have access to a suitable device.

This mobile executive can also expect to start to see a convergence of accessories. At present it is possible to purchase mobile phones with calculator and clock functionality. With the emergence of electronic cash technologies, the traditional wallet will start to disappear. The combination of wireless and smart cards will provide an easy to use and safe method for creating and storing encryption keys and other sensitive information. Finally, WAP-based applications will reduce the need to carry a PDA or organizer.

If we now look at the main issues facing the mobile commerce developer, we can build on the points already covered.

As we have said, it is good practice not to ask a user to enter data already known about that customer by the financial services organization. This is particularly true for the WAP phone, as data entry via a keypad will be slow. It is also likely that the user will be on the move and access to personal data such as share portfolios will be limited. Therefore, data entry should be minimized using the following techniques.

▸ Personalization based on the customer profile or previous usage of the application can limit the navigation required by presenting menus in an optimized order.

- ▸ Default values for data fields should be offered where possible.

- ▸ Using an associated Web site, it should be possible to create lists, which can then be accessed via the WAP phone (e.g., shares to track).

Attention has already been drawn to the more sophisticated aspects of Web development such as cookies, browser-specific server-generated responses, and the WAP alternative of variables defined outside the scope of any particular WML card. However, there is one significant disadvantage to this for the developer of financial services applications. These variables are not Web site-specific and also do not time out. It is conceivable that a user could move onto a second Web site, which innocently or maliciously accessed these variables. It is therefore important that no customer-specific information, or anything that could compromise security, is stored in these variables. The Web site should clear out or delete the variables as part of the process of logging off the WAP site, but this cannot be relied upon, as the user may simply disconnect during the session, or there may be a loss of network service.

Adopting the fifth principle of distributed computing, or the more colloquial version "anything that can go wrong will, usually at the worst possible time," it is obviously vital that error handling must be watertight for WAP applications, which execute financial transactions. The techniques for building financially secure Web application servers are now well documented [5]. However, the loss of connection to the client after a transaction is committed and the ability to recover from this are usually omitted from the normal Web application design. Most financial host systems do not usually have the ability to work 100% in real time. Hence, a funds transfer followed immediately by a mini-statement will probably not show the transfer. It is therefore important for the mobile user to have some other form of positive notification that the transaction did indeed take place. One way to implement this would be for the WAP client to interact with the WAP application server to indicate that a confirmation notification has been received and displayed. Failure to receive this notification, presumably due to a loss of service, would cause the server application to send a confirmation via SMS or to have the information prominently displayed when the user reconnects.

WAP-based security is covered specifically in Chapter 7. However, for the financial services organization, security must always be given the highest attention.

The three cornerstones of authentication are well documented. Currently, the WAP phone can implement two of the three. The first is asking for information private to the customer. This is in the form of one or more PINs. The second is physical possession of something known to belong to the customer which cannot easily be copied; for example, in the mobile phone it is the SIM card. The third cornerstone, which will be available soon, is the measurement and validation of a physical characteristic of the customer. For the mobile phone this is likely to be voice recognition.

Using the above techniques, the GSM phone is as secure as is practically required to execute personal financial transactions. In addition, the SIM card is a great container for a personal authentication key and encryption algorithm. It is virtually impossible to copy and can be protected by its own PIN. It is likely therefore that the mobile phone will become a de facto standard for personal e-commerce. Early evidence of this can already be seen as secure messaging to mobile phones is being used to confirm larger transactions executed on other channels such as via a call center.

10.15   Sample banking application

The following figures show a hypothetical WAP application. The sample screens were designed and constructed by AIT Ltd. The first application has been designed to run on a low form factor device, which is likely to represent the low to medium end of the WAP phone market. Figure 10.15 shows some sample screens from a simple banking application.

As the form factor of the device increases, it is possible to increase the amount of information displayed at any one time on the device. Figure 10.16 shows a product selector screen from the imaginary Henley Bank.

Using a little poetic license, it is possible to imagine what a more sophisticated mobile financial services application may one day look like. See Figure 10.17.

10.16   Possible mobile financial services applications using WAP

It is worth examining some of the financial services offerings that are possible using WAP and SMS-based messaging. Using push technology (see

Figure 10.15 Simple WAP banking application (pictures courtesy of AIT Ltd.); shows (a) the welcome page, (b) the product selector, (c) the account balance, and (d) the funds transfer.

Figure 10.16 Sample WAP banking application (courtesy of AIT and Ericsson).

Chapter 6 for more details about the concept of push services and its implementation in WAP), it is possible to send messages to indicate that an event has been triggered. These events could fall into one of three categories.

1. Subscription based, such as financial news feeds;

2. User defined, such as defined stock movements (e.g., a +/−2% change in BT shares);

Figure 10.17   Sample sophisticated WAP financial services application (picture courtesy of AIT and Ericsson).

3. Financial service provider defined, such as a large transaction has hit your account, you have exceeded your credit card limit, or your mortgage is ready for you to exchange contracts on a house.

It is also likely that the user will want to browse WAP-based financial information portals looking for general news and market information. These will include FTSE indexes, interest and exchange rates, etc. The ability to execute simple transactions with a variety of financial institutions will include home banking functionality, share dealing, and possibly foreign currency transactions.

As sophisticated applications are developed, it may be possible to obtain illustrations or quotations to aid in purchasing new financial products. In fact, it will then be feasible to get impartial best-value quotes from financial intermediary sites. These services will be geared to the needs of the mobile customer (e.g., "I am in a boat showroom; get the best price for marine insurance.").

It will also be possible to offer services based on the current location of the customer. These will include simple informational services such as where the nearest ATM or branch is. More sophisticated applications would include emergency credit card or traveler's check replacement.

In all cases, it is important to understand the circumstances of the mobile customer, who may, for example, be in a taxi or airport departure lounge; hence, the services must be executed simply and quickly.

10.17   The role of other service delivery channels

It is important to remember that the WAP phone is likely to be just one of several channels supported by the financial services organization. By taking a synergistic approach to the implementation of services, it is possible to augment the mobile applications by appropriate use of the other channels, in particular, voice, call center, and the Internet.

The Internet can be used to select and configure WAP-based services that would be harder and more time-consuming to do on the phone itself. This could include selecting shares to be monitored, or entering service preferences. The combination of WAP and Web is a very powerful combination for the next generation of financial services-based portals.

Capitalizing on the voice capabilities of the WAP phone, there may be situations where it may be appropriate to hand off calls to an IVR unit. This can be done securely, and therefore there is no need for the IVR to reauthenticate the user. These facilities could be used to semiautomate operations which have not been specifically implemented in the WAP-based application. For example, a WAP menu to leave a voice message could be used to request the reissue of a statement or order a new checkbook.

Operator-assisted Internet sites are now becoming more common, where the user is connected to a call center agent via a "call me" facility on the Web page. This can be equally applicable to the WAP user, and can operate seamlessly with the phone user. The WAP technology behind such services is the WTA standard, which is discussed in detail in Chapter 4.

10.18 The personal mobile phone and customer relation management

It is possible to implement all the standard customer relation management (CRM) techniques [6] (e.g., campaign-based and one-to-one marketing on the mobile channel). However, care must be taken not to alienate the customer with inappropriate push-based techniques. The personal nature and limited display capabilities of the device means that a customer is more likely to read an inappropriate (to him or her) marketing message on the mobile phone, than if it were to appear in the inbox of a PC-based mail reader. Also, it is currently impossible to filter out these spam or junk mail messages on the phone.

One safeguard against this would be to allow the customer to set his or her own "intrusion level." This is also likely to change based on the time of day, in the same way that intelligent networks will allow all calls from the office to be routed to voice mail after 7 P.M. Rather than these intrusion levels being prescriptively defined in advance of using the service, a more adaptive approach could be taken. At the end of a message could be appended "Are you happy to receive these messages, all the time, during office hours, or never?"

The biggest challenge to confront the marketing department of the mobile financial services organization is who owns the customer. The market forces that will determine this are similar to those of the Internet or iTV—in the final analysis it will be based on who is adding most value to a given service.

10.19   Next generation of WAP-based financial services and applications

The next generation of WAP-based financial services and applications will be driven by an increase in network bandwidth using technologies such as the general packet radio service (GPRS) and high-speed circuit switched data (HSCSD). Announcements from the network providers have already been made. Another factor will, of course, be the increased sophistication of the mobile devices of the future. More advanced phone and server-based speech recognition will eliminate the need to use the keypad.

The next paradigm shift, however, will probably come with the combination of the introduction of e-cash and Bluetooth wireless communications. In this world it will be possible for your phone to automatically pay for road and parking tolls without manual intervention. The credit balance on the phone could also be automatically topped up via a call to the appropriate bank.

10.20   Conclusion

We are about to enter a new era of mobile commerce, and as much as any financial services application could be described as exciting, this will be exciting. The very early days will be frustrating, and many people will spend a great deal of effort exploring technological blind alleys. However, to paraphrase those now famous words, WAP will be one small step for technology and a giant leap for mobile commerce.

References

[1]   See http://www.gartnerweb.com.

[2]   See http://www.microsoft.com/PressPass/press/1999/May99/WAPpr.htm.

[3]   See http://www.bluetooth.com.

[4]   Eckstein, R., *XML Pocket Reference Extensible Markup Language*, 1st Ed., Sebastopol, CA: O'Reilly & Associates, 1999.

[5]   Garfinkel, S., and G. Spafford, *Web Security & Commerce*, Sebastopol, CA: O'Reilly & Associates, 1997.

[6]   Petersen, G. S., *Customer Relationship Management Systems: ROI and Results Measurement*, Downers Grove, IL: Strategic Sales Performance, Inc., 1999.

Acronyms

AES   advanced encryption standard

CA   certification authority

CC/PP   composite capabilities/preference profiles

CDF   channel definition format

CDPD   cellular digital packet data

CGI   common gateway interface

CRL   certificate revocation list

CSD   circuit switched data

DES   data encryption standard

DLP   discrete logarithm problem

DSA   digital signature algorithm

DTD   document type definition

ECDLP   elliptic curve discrete logarithm problem

GPRS   general packet radio service

GUI   graphical user interface

HDML   handheld device markup language

HLR   home location register

HMAC   hashed machine authentication code

HTML   hypertext markup language

HTTP   hypertext transport protocol

HTTPS   hypertext transport protocol secure

IAP   Internet access provider

IDEA   international data encryption algorithm

IETF   Internet Engineering Task Force

IFP   integer factorization problem

IMAP4   interactive mail access protocol

INAP   intelligent network application part

IPsec   Internet protocol secure

ISP   Internet service provider

ITTP   intelligent terminal transfer protocol

ITU   International Telecommunications Union

LDAP   lightweight directory access protocol

MAC   message authentication code

MD5   message digest algorithm 5 (designed by Ron Rivest)

MSISDN   mobile station international subscriber directory number

NIT   network integration test

NMS   network management system

OTA   over the air

PDA   personal digital assistant

PKI   public key infrastructure

PKIX   public key infrastructure for the Internet (designed by the IETF)

POP3   post office protocol

RAS   remote access server

RC4   Rivest Cipher 5

RSA   Rivest Shamir Adleman scheme

S/MIME   secure/multipurpose Internet mail extensions

SAP   service access points

SAT   SIM application toolkit

SHA-1   secure hashing algorithm 1

SIA   session initiation application

SIM   subscriber identity module

SM   short messages

SMSC   short message service center

SMTP   simple mail transfer protocol

SNMP   simple network management protocol

SSL   secure socket layer

TCP/IP   transport control protocol/Internet protocol

TLS   transport layer security

TTML   tagged text markup language

TUI   telephonic user interface

UA   user agent

UAProf   user agent profile

UDP   user datagram protocol

URI   uniform resource identifier

URL   uniform resource locator

USSD   unstructured supplementary service data

VPN   virtual private network

WAE   wireless application environment

WAR   wireless application reader

WBMP   wireless bitmap

WBXML   wireless binary XML

WDP   wireless datagram protocol

WIM   wireless identity module

WML   wireless markup language

WSP   wireless session protocol

WTA   wireless telephony application

WTAI   wireless telephony application interface

WTAS   wireless telephony application servers

WTLS   wireless transport layer security

WTP   wireless transaction protocol

WWW   World Wide Web

XML   extensible markup language

About the Authors

Simon Blake-Wilson is a cryptographic mathematician at Certicom Corp. He is involved in a number of standards efforts, including elliptic curve initiatives in ANSI, IEEE, ISO, and SECG. His research interests include secure protocol design, public-key infrastructures, and security management. Mr. Blake-Wilson holds a Ph.D. in mathematics from the University of London, and is currently a part-time visiting researcher at the Centre of Applied Cryptography at the University of Waterloo, Canada. He can be reached at: sblakewilson@certicom.com.

David Brazier has wide experience in systems development, with an emphasis on user interface design. He was previously employed at PA Consulting and Logica, where he worked with users of many different computer systems—air traffic controllers, postmasters, and financial traders—helping them design the systems they need to achieve their goals. Most recently, Mr. Brazier has been working with the Scottish Health Service to develop an integrated system for general practice doctors and their staff, which is now in use in hundreds of practices across the country.

Martin Frost is the WAP development leader at Dynamical Systems Research and has extensive experience in the development of WAP browsers, WML encoders, and WMLScript compilers and interpreters as well as the development of advanced WAP applications.

Robert Gallant is a product development manager at Certicom Corp. He recently led the team that developed Certicom's WTLS implementation.

Mr. Gallant is also an active researcher and is well known for his work on efficient methods for implementing elliptic curve cryptography. He holds a Ph.D. in mathematics from the University of Waterloo, Canada. He can be reached at: rgallant@certicom.com.

Jarno Haikonen currently works for Tecnomen as a product manager responsible mainly for wireless Internet-related products. His work includes defining products and services based on market needs and trends. Mr. Haikonen is also studying telecommunication management at the Helsinki University of Technology.

Ian Hosking has more than 10 years of experience in mobile computing and wireless data applications. Before joining Digital Mobility, he was employed at Logica, where he worked alongside blue-chip financial and telecommunications organizations to define effective mobile computing solutions. At Logica he was also responsible for the realization of the world's first Internet photojournalism system, as used by climber Chris Bonington to update the his Web site during a successful climb of unexplored Tibetan peaks in 1997.

Kai Kauto holds an M.Sc. in engineering from Oulu University, Finland. He joined Tecnomen in 1997 as product manager for Internet products. Currently, he is the director of product management and product marketing and a member of Tecnomen's board of directors. Previously, he worked as an export sales manager in a Nordic IT company called WM-data.

Bo Larsson has been involved in WAP standardization activities on behalf of Ericsson since 1997. He has been a member of the "push core team"—the team actively leading the push effort in the WAP Forum —since it was formed in 1988.

Magnus Larsson holds an M.Sc. in electrical engineering and works as a consultant at AU-System. He has participated in the WAP Forum work with WTA standardization on behalf of Ericsson since 1998.

Janet Loughran has a B.Sc. (Hons) in microelectronic systems from the University of Ulster. Ms. Loughran has worked in the telecommunications industry since 1987, and for the past 9 years has developed control software for a number of mobile standards such as GSM, DECT, TETRA, and PDC. She currently works with APiON Ltd. as a presales technical consultant. She can be reached at: wapbdev@apion-tss.com.

Hugh MacDonald is a cryptographic consultant for Certicom Corp. He recently moved to Vancouver after several years of working for Certicom in the Toronto area. Mr. MacDonald holds a B.A. in mathematics from the University of Waterloo, Canada. He can be reached at: hmacdonald @certicom.com.

Stuart Marsden was a director of AIT from 1993 to 1999, dedicated to providing multichannel customer-engagement applications focused on financial services applications. For the past 3 years, Mr. Marsden has focused on emerging technology for new channels and e-commerce. Recently, he established Et al Innovations Ltd., whose mission is to apply innovative approaches to exploit emerging technology for delivering business benefits with a focus on e-commerce and knowledge management. He can be contacted at: stuart@et-al.co.uk.

Per Ocklind is responsible for WAP technology marketing within the Ericsson Product Unit Wireless Internet Applications. He joined Ericsson in 1989, working as product manager for radio modems and terminals for Mobitex. In 1996 he transferred to Ericsson Radio Systems (GSM) where he functioned as a "human interface" to the terminal side. Since 1997 he has been involved in the creation and administration of WAP.

Prakash Panjwani is the director for wireless market development and product marketing at Certicom Corp. He has been an active participant in wireless standards for several years, having served as the chair for CDMA signaling task group within TIA while he was with Motorola's cellular division. Mr. Panjwani holds a B.S. in electrical engineering from Columbia University and an M.S. in information networking from Carnegie Mellon University. He can be reached at: ppanjwani@certicom.com.

Johann Reindl has more than 5 years of experience in the telecommunications industry. After his MBA he worked as the product marketing manager for intelligent network services at GPT Ltd. in the United Kingdom. Since then he has been with T-Mobil, Deutsche Telekom's mobile operator. After being involved in a number of different assignments in corporate strategy and marketing, he is currently working as project manager in product marketing and has been heavily involved in the commercial introduction of WAP services.

Greg Sigel manages business development in the wireless and international markets for Certicom Corp. He has initiated and been a key driver of Certicom's involvement in WAP and other cellular standards bodies and forums. Mr. Sigel holds an MBA in marketing, finance, and international business from the J. L. Kellogg Graduate School of Management at Northwestern University. He can be reached at: gsigel@certicom.com.

Marcus Taylor is the director of products and services at Digital Mobility Ltd. Prior to Digital Mobility, he worked for Logica as a managing consultant leading the Mobile Innovation Group in Cambridge, United Kingdom, and then in Stockholm, Sweden. Prior to those positions, Mr. Taylor worked for IBM as a human factors specialist in various locations around Europe. He has extensive experience in the application of business and human factors methods and techniques to applications in electronic/mobile commerce, mobile computing, new media, and the Internet and intranet.

Ilkka Teppo is a product manager for wireless Internet products at Tecnomen. Before joining Tecnomen, Mr. Teppo was employed by Radiolinjia, a large Finnish-based mobile operator. He holds an M.Sc. from the Helsinki University of Technology. He is currently concentrating his team's efforts on extending Tecnomen's wireless Internet portfolio to include a wider range of innovative wireless Internet and e-commerce services.

Marcel van der Heijden has a Ph.D. in nonlinear dynamical systems and neural networks and has been involved in GSM value-added services since 1996 and WAP software and applications development projects since 1998.

Index

Recent Titles in the Artech House Mobile Communications Series

John Walker, Series Editor

Wideband CDMA for Third Generation Mobile Communications, Tero Ojanperä and Ramjee Prasad, editors

Wireless Communications in Developing Countries: Cellular and Satellite Systems, Rachael E. Schwartz

Wireless Technician's Handbook, Andrew Miceli

For further information on these and other Artech House titles, including previously considered out-of-print books now available through our In-Print-Forever® (IPF®) program, contact:

Artech House
685 Canton Street
Norwood, MA 02062
Phone: 781-769-9750
Fax: 781-769-6334
e-mail: artech@artechhouse.com

Artech House
46 Gillingham Street
London SW1V 1AH UK
Phone: +44 (0)20 7596-8750
Fax: +44 (0)20 7630-0166
e-mail: artech-uk@artechhouse.com

Find us on the World Wide Web at:
www.artechhouse.com